The Hydrogen Energy Transition: Moving Toward the Post Petroleum Age in Transportation

The Hydrogen Energy Transition: Moving Toward the Post Petroleum Age in Transportation

Edited by

Daniel Sperling and James S. Cannon

AMSTERDAM • BOSTON • HEIDELBERG • LONDON • NEW YORK • OXFORD
PARIS • SAN DIEGO • SAN FRANCISCO • SINGAPORE • SYDNEY • TOKYO

ELSEVIER
ACADEMIC
PRESS

200 Wheeler Road, 6th Floor, Burlington, MA 01803, USA
525 B Street, Suite 1900, San Diego, California 92101-4495, USA
84 Theobald's Road, London WC1X 8RR, UK

This book is printed on acid-free paper. 

Library of Congress Cataloging-in-Publication Data
Application submitted

British Library Cataloguing in Publication Data
A catalogue record for this book is available from the British Library

ISBN: 0-12-656881-2

For all information on all Academic Press publications
visit our website at www.academicpress.com

Printed in the United States of America
04 05 06 07 08 09 9 8 7 6 5 4 3 2 1

Dedication

We dedicate this book to Barry McNutt. Barry had been a central presence at the biennial Asilomar conferences on transportation and energy since their inception 15 years ago. He originated the "open mike" session, presented many insightful talks, and served on many conference organizing committees. It was Barry's proposal that we devote the 2003 conference to hydrogen. He served as conference co-organizer and co-authored a chapter. Barry was 57 years old when he died on Sunday, November 16 at his home in Arlington, Virginia, after a long and courageous battle with cancer.

Congressman W.J. "Billy" Tauzin of Louisiana honored Barry in the *Congressional Record* on December 8, 2003. He noted that Barry was a distinguished energy policy analyst and outstanding public servant who "...was a strong advocate for increasing domestic energy supplies and improving energy efficiency. While we may have disagreed with Barry's analysis at times, we always respected it because we knew it was coming from a man with great intellectual gifts and unblemished integrity. Barry McNutt was only interested in good policy not politics, but he recognized that good policy happens through the legislative process. He worked tirelessly to formulate policy options that informed and enlightened the process ... Barry often told his colleagues that the most important thing is to produce solid analysis that will stand the test of time and he did that with talent and great care."

We honor Barry as a longtime friend and colleague, who will forever inspire us.

Acknowledgements

This book demonstrates the huge interest in hydrogen and the willingness of so many people and organizations to provide time, money, and talent in exploring this topic. The book is based on the IX Biennial Asilomar Conference on Transportation and Energy, held July 29 to August 1, 2003 in Pacific Grove, California. The book comprises a selection of papers presented at the conference, plus two additional papers prepared after the meeting by conference participants.

The conference would not have been possible without the generous support of several organizations. Most generous was the U.S. Department of Energy. Other major sponsors were the Ministry of Natural Resources Canada and University of California Transportation Center, with additional support provided by the William & Flora Hewlett Foundation, Energy Foundation, U.S. Federal Transit Administration, and Calstart.

The conference was hosted and organized by the Institute of Transportation Studies at the University of California, Davis (ITS-Davis), under the auspices of the US National Research Council's Transportation Research Board – in particular, the standing committees on Energy and Alternative Fuels.

The conference program was directed by Barry McNutt, to whom this book is dedicated, along with David Greene, Jack Johnston, Marianne Mintz, Peter Reilly-Roe, Phil Patterson, Dan Santini, Lee Schipper, and Dan Sperling. This committee worked closely in crafting a set of speakers and topics that was engaging and insightful.

In addition to the many authors, we want to acknowledge the generous assistance of a large set of peer reviewers, drawn from attendees at the conference, who provided valuable feedback and suggestions to authors. Book production was assisted by Jeff Georgeson, who provided copy editing assistance. We especially appreciate the efforts of Christine Minihane of Elsevier, who shepherded and supported the entire book project from its inception.

We especially want to acknowledge the many attendees of the conference listed in Appendix B. These invited leaders and experts, coming from many parts of the world and many segments of society, enriched the conference with their deep insights and rich experiences.

Contents

1. Introduction and Overview by Daniel Sperling and James S. Cannon 1

2. Back from the Future: To Build Strategies Taking Us to a Hydrogen Age by David Sanborn Scott 21

3. Prospecting the Future for Hydrogen Fuel Cell Vehicle Markets by Kenneth S. Kurani, Thomas S. Turrentine, Reid R. Heffner, and Christopher Congleton 33

4. Fuel Cell Hybrid Vehicles: The Challenge for the Future by Taiyo Kawai 59

5. Where Will the Hydrogen Come From? System Considerations and Hydrogen Supply by Joan Ogden 73

6. Clean Hydrogen from Coal with CO_2 Capture and Sequestration by Richard D. Doctor and John C. Molburg 93

7. Doing Good by Doing Well: Entrepreneurship in the Hydrogen Transition by David L. Bodde 105

8. Hydrogen from Electrolysis by Chip Schroeder 121

9. The President's U.S. Hydrogen Initiative by Steve Chalk and Lauren Inouye 135

10. The Hydrogen Transition: A California Perspective by James D. Boyd 147

11. U.S. Hydrogen Activities—A European Perspective by Barend van Engelenburg 155

12. Lessons Learned from 15 years of Alternative Fuels Experience—1988 to 2003 by Barry McNutt and David Rodgers 165

13. Lessons Learned in the Deployment of Alternative Fueled Vehicles by Bernard I. Robertson and Loren K. Beard 181

14. Understanding the Transition to New Fuels and Vehicles: Lessons Learned from Analysis and Experience of Alternative Fuel and Hybrid Vehicles by Paul Leiby and Jonathan Rubin 191

15. The "Chicken or Egg" Problem Writ Large: Why a Hydrogen Fuel Cell Focus is Premature by John M. DeCicco **213**

16. The Case for Battery Electric Vehicles by Paul B. MacCready **227**

17. Hydrogen Hope or Hype by Daniel Sperling and James S. Cannon **235**

Appendix A About the Editors and Authors **241**

Appendix B Asilomar Attendee List 2003 **249**

Index **255**

CHAPTER 1

Introduction and Overview

Daniel Sperling, Institute of Transportation Studies, University of California, Davis, and James S. Cannon, Energy Futures, Inc.

In 2003, U.S. president George W. Bush, European Union president Romano Prodi, and California governor Arnold Schwarzenegger all endorsed the hydrogen energy economy. Many papers and books provide details of what many believe is an imminent future. But is it imminent, and what exactly are we imagining? Transforming today's oil dominated transportation system to one running on hydrogen is a daunting challenge, perhaps the most important facing the world today.

The chapters in this book present strategies, policies, and technologies that can nudge transportation and other systems away from oil and toward hydrogen. They are written by participants in the 9th Biennial Asilomar Conference on Transportation and Energy, held in California in the summer of 2003.

The conference theme at Asilomar was "The Hydrogen Transition." Most conference presenters, though not all, concur with Bush, Prodi, and Schwarzenegger in endorsing the hydrogen vision. They believe that hydrogen will emerge in the twenty-first century as the primary transportation energy fuel. The more difficult and contentious questions are how and when to get there, and what energy resources to use. This book does not define a specific pathway to hydrogen. The goal is to examine key issues in the transition, inform the public debate, and suggest strategies for how to begin to get from here to there. As a practical guide, the book's starting point is the transportation world of today, where the overpowering inertia of the oil, automotive, and electric power industries discourages change.

The Daunting Challenge of Hydrogen Transportation

Most oil is used for transportation; two-thirds of all oil used in the United States and half of all oil used worldwide is for transportation. Thus, this book focuses on the transport sector, where oil use continues to increase and energy security problems are most acute. Transforming the global transportation energy economy from oil dependent vehicles of the twentieth century to sustainable hydrogen systems will be the greatest human technological achievement since ... well, maybe ever. The first problem is that free hydrogen does not exist naturally on Earth in its gaseous form. Much lighter than air, any free hydrogen produced by natural processes during the planet's history has long since floated into space or been captured in more complex molecules.

Hydrogen, therefore, is more like effervescent electricity than naturally occurring oil. It is an energy carrier. Like electricity, hydrogen can be produced from a wide range of common materials, including water, natural gas, and coal. Once formed, gaseous hydrogen—which consists of two bound hydrogen atoms—can be burned as a fuel or used in a fuel cell to generate electricity.

The key to the hydrogen economy and the upsurge in interest in hydrogen in the past few years is the rapid development of fuel cell technology. Fuel cells are viewed as the holy grail by the automotive industry. They are much more energy efficient than internal combustion engines, produce no pollution, and, because they are well suited to hydrogen, facilitate a shift away from petroleum. But the attractions extend beyond just energy and environmental benefits—and that may be the key.

The use of hydrogen in fuel cell vehicles (FCVs) entails the electrification of the automobile's power train, an important attraction to automakers due to the high torque and power output of electric motors. Since the first power station was built in 1882, most everyday appliances—from toasters to toothbrushes—have switched to electricity. Motor vehicle drivetrains have thus far remained immune, although a slew of electric systems, such as power steering, brakes, and windows, now create a large parasitic electric load, draining power from the engine's electric generator and forcing the introduction of higher capacity components.

Drivetrain electrification has failed to date mostly because of the difficulties in providing the electricity. Either it must be stored onboard, which is difficult and expensive to do, or large amounts of electricity must be generated onboard, also difficult and expensive. The car, therefore, remains one of the last great frontiers for electricity to conquer, and hydrogen fuel cells appear to be the way to do this.

The relationship and analogies of hydrogen and electricity are central to this book. As suggested above, their future is inextricably linked. They will be competitors, and they will be partners. Once formed, electricity and hydrogen are equal opportunity energy forms, often called "currencies" because of their similarities to money. Each can be produced or "earned" from a range of resources and used or "spent" to power any electric appliance in the world. The advent of hydrogen as the energy source for FCVs completes the electrification revolution by turning the motor vehicle into a viable electric car.

Hydrogen and electricity create a unified energy system, and facilitate the expansion of distributed energy systems, a trend taking hold around the world. For instance, when electricity is not needed to run a FCV—most cars sit idle more than 90 percent of each day—its fuel cells can still be used to generate electricity, feeding it back to power lines for use in homes and businesses elsewhere. This promise of small-scale pollution-free local electric power production, termed "distributed" generation, runs counter to the twentieth century model of building large generating plants far from population centers and then connecting generation sites to consumers through a far reaching maze of power lines. To make distributed power a market reality, the electric power industry will have to reform and restructure itself significantly—a huge task resisted by many large electric utility companies.

In the process of electrifying the automobile, the hydrogen FCV has no need for the internal combustion engine (ICE) or the oil it burns. Since their inventions in the late 1800s, spark ignition ICEs used to burn gasoline and compression ignition ICEs used to burn diesel have powered virtually all the billion plus motor vehicles built to date. ICEs are installed in virtually all the 50 million cars, trucks, and buses built annually around the world today.

Roughly a hundred years ago, oil fuels and their ICEs challenged the horse drawn transportation era. The "horseless carriage" offered in 1900 was expensive and primitive. It prevailed ultimately because it was a better product at a time when the world was desperate for one. The automobile didn't require enormous chunks of agricultural real estate to produce its fuel—roughly one-quarter of all U.S. cropland in 1900 was used to grow hay for horses—and it was faster and cleaner. At one horsepower, the power output of a horse was small, restricting travel to a plodding few miles per day. The resulting pollution—manure and urine—created one of the major urban public health concerns of the day (Cannon, 1995).

Even so, the victory of motor vehicles over the horse was not quick or easy. The 20 million horses used for transportation in the United States in 1900 swamped by a factor of 10,000 the few thousand motorized vehicles on the road at the time. The gasoline car was not even the leading competitor to the horse. They shared the motorized vehicle market with battery electric and steam powered vehicles (Cannon, 1995). Despite the advantages of motorized vehicles, opposition was strong because the new fuels and technologies were disruptive. Very few farmers became oil drillers and even fewer stable boys became automotive engineers. Hundreds of thousands of jobs were lost in the transition.

The motor vehicle transition required an enormous automotive engineering effort, development of many new technologies, and adoption of an entirely new method of manufacturing. It was not until 1925 that the number of registered automobiles exceeded the number of horses that had been used for transportation in 1900, though many automotive and oil fortunes were earned long before then.

The advent of the motor vehicle, a long shot in 1900, is now complete. With continuing improvements—more power, reliability, and comfort—the superiority of motor vehicles over horses is well established. More than 750 million gasoline- or diesel-fueled vehicles now course the world's roadways, nearly one-third of which are in the United States (Davis and Diegel, 2003). Roughly 100 new cars or trucks are built worldwide every minute, nearly every one equipped with an ICE and fueled by oil. The global automotive industry receives a trillion dollars per year in revenue and spends tens of billions of dollars annually on research, almost all of it for minor enhancements to the current technology. Fuel cell and hydrogen research and development (R&D) are receiving in the early years of the decade roughly a billion dollars a year from the automotive industry, a substantial amount but only about one percent of what it spends on other R&D. Much more would need to be spent to launch a transition to hydrogen fuel cells. Other companies would also play a role, but only 10 to 15 global automakers have the skills and resources to make this happen. When and under what circumstances would the industry be willing to make the large investments in this new and disruptive technology? More to the point, are companies ready and willing to take the risk, and what public policy would be needed to encourage them to do so?

A century ago, the car would not have succeeded without a new, cheap, and powerful fuel in the form of gasoline. Refined from oil, itself discovered only in 1859, gasoline was first an unwanted by-product of the production of kerosene, in demand then as a lighting fuel. Sometimes discharged into rivers as a waste from oil refining, gasoline turned out to be a great fuel for ICE powered automobiles. With very high energy content per volume, gasoline contains much more energy than an equal weight or volume of hay, the transportation fuel it replaced. Huge investments in gasoline production and distribution were within the financial reach of oil companies. The gasoline fueling infrastructure replaced the hay producing agricultural industry and proceeded to expand in step with automotive manufacturing.

During the past century, nearly one trillion barrels of oil have been extracted from the ground worldwide, most of it refined to fuel motor vehicles. The multibillion dollar global fleet of oil supertankers, the millions of miles of fuel pipelines, and the hundreds of thousands of vehicle fueling stations dotting the landscape are testaments to the rise of the oil industry in service of the automobile. Little of this investment is likely to be particularly useful in a hydrogen transportation energy economy. The transition to hydrogen will require a transformation or replacement of the current automotive and oil industries. The future will be populated by fuels and technologies not available today.

Energy Security and the Environment

The soaring appetite for oil transportation fuels in the United States and elsewhere is creating tremendous tensions and risks. It is widely accepted that a transition away from petroleum and fossil fuels must occur within decades rather than centuries.

The focus in this book is the United States. It is by far the largest consumer and importer of oil, the largest producer of greenhouse gases, and the largest economy. The almost 220 million cars, buses, and trucks traveling U.S. roads are the main reason for the country's steadily rising reliance on foreign oil. As shown in Fig. 1-1, this reliance has grown from 35 percent at the time of the 1973 oil embargo to 60 percent in 2003.

As the transportation system's demand for oil has grown, U.S. oil production has declined. Since its historical peak in 1970, U.S. oil production has dropped more than 25 percent, and since 1986, oil use in transportation alone has exceeded the total industry's domestic production. This decline is charted in Fig. 1-2.

Meanwhile, world oil production grew by 14.5 percent from 1992 to 2001, from 65.2 to 74.7 million barrels a day, and world competition for these global oil supplies has steadily escalated. Much of the developing world is now aspiring to replicate these gasoline- and diesel-dependent transportation systems, especially in China, where 30 percent of its oil is now imported, a remarkable increase from zero in 1993 (EIA, 2003b). China is now behind only the United States in its projected growth in oil imports during the next decade.

In the period from 2000 to 2020, world oil use is likely to exceed the amount used in the entire preceding industrial era. A U.S. Central Intelligence Agency report, *Global Trends: 2015*, has concluded that economic and population growth are expected to increase oil demand

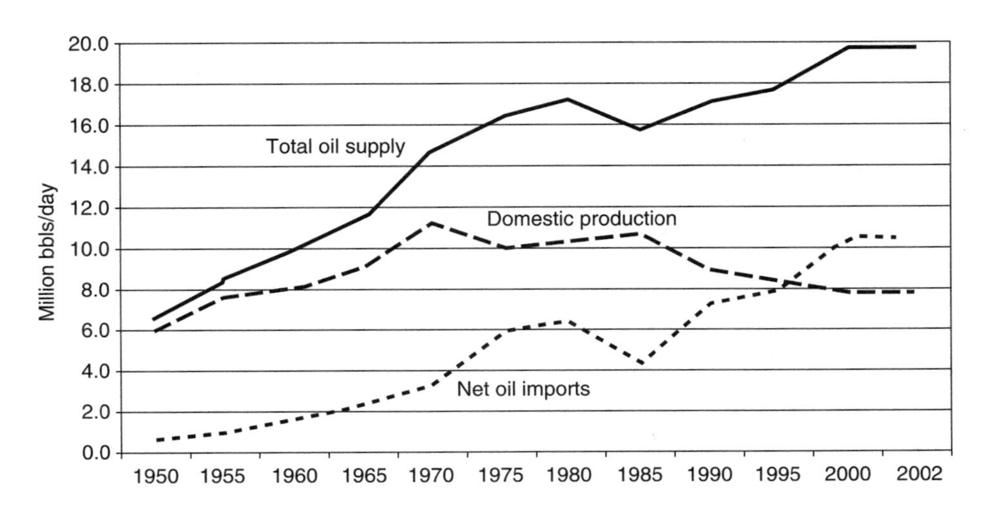

FIGURE 1-1. Growth in U.S. Net Oil Imports: 1950-2002. (*Source:* EIA 2003a)

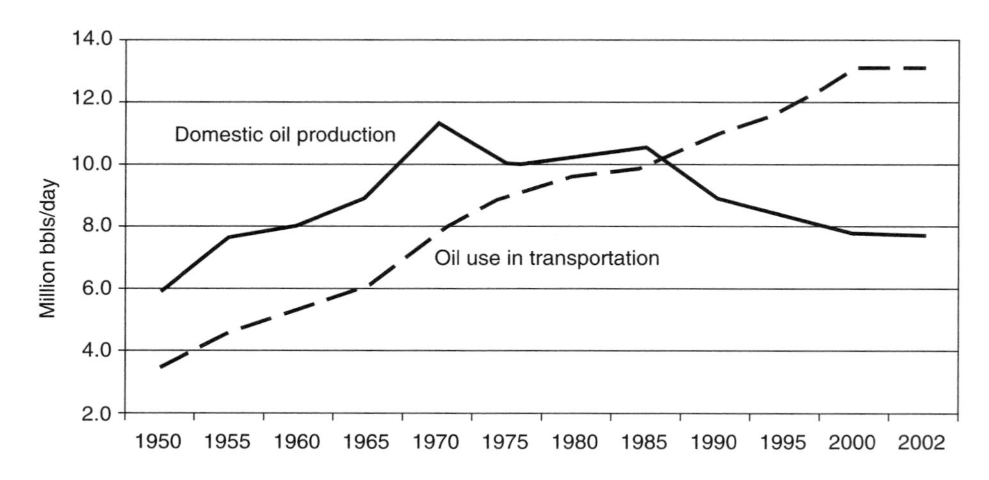

FIGURE 1-2. U.S. Oil Production and Domestic Use: 1950-2002. (*Source:* EIA 2003a)

another 50 million barrels per day to more than 120 million barrels in 2015 (CIA, 2001). The increase alone is almost as large as the current production by nations within the Organization of Petroleum Exporting Countries (OPEC). Asia is expected to drive the expansion in energy demand, replacing North America as the leading energy consuming region and accounting for more than half of the world's total increase in demand between 2000 and 2015. The study's most startling conclusion is that by 2015 only one-tenth of Persian Gulf oil will be directed to western markets; 75 percent will go to Asia.

The economic repercussions of oil dependency are substantial. Oil imports drained over $100 billion from the U.S. economy in 2003, or roughly $400 per person. Thousands of jobs have been sent abroad to those who produce and deliver oil to U.S. shores. And expensive U.S. foreign policy and military actions to secure low priced oil from abroad cost consumers many billions of dollars. Perhaps even more troubling is the risk of supply disruptions—easily imaginable as dependence on the politically unstable Persian Gulf escalates.

The problems of oil dependence do not stop with access to supply. Oil burned in motor vehicles causes roughly half the air pollution in most U.S. cities. Smog (mainly ozone pollution caused by vehicle emissions) and small particles (especially from diesel exhausts) inhaled deeply into lungs cause asthma and widespread respiratory distress. The U.S. Environmental Protection Agency report *Latest Findings on National Air Quality: 2002 Status and Trends* concludes that a significant portion of the U.S. population, roughly 146 million out of 260 million people, still live in areas where health-based air quality standards are being violated. Air quality is improving in virtually all regions but remains a serious health hazard in many inner cities and regions with meteorological temperature inversions (such as Los Angeles).

The Asilomar Conference

The promise of a zero-polluting, sustainable-energy system provides the "pull" toward a new transportation paradigm. The race to find alternatives to depleting oil resources and increasing carbon dioxide emissions, as well as persistent urban air pollution, are the key "pushes" for change in the current system. The question addressed at Asilomar was how to launch viable hydrogen and FCV industries. The transition will be hampered by the huge commitment to the status quo, not only by automotive and energy industries but also by entangling rules, permits, and policies that discourage change.

From July 29 through August 1, 2003, over 200 energy, environmental, and transportation policy experts traveled to the Asilomar Conference Center in Pacific Grove, California, where they grappled with the complexities of transforming the U.S. transportation energy system. The conference was the ninth in a series begun in the 1980s to discuss the pressing transportation energy issues of the day. Since 1991, the event has been held every two years. The conference is hosted by the Institute of Transportation Studies of the University of California, Davis (ITS-Davis), on behalf of the U.S. Transportation Research Board. The principal sponsor for the 2003 event was the U.S. Department of Energy (DOE). Other sponsors were Natural Resources Canada, University of California Transportation Center, U.S. Federal Transit Administration, Calstart, Energy Foundation, and William and Flora Hewlett Foundation.

The sessions at Asilomar attempted to untangle the web of political, business, consumer, and public interest considerations raised by a transition to hydrogen. Among questions posed to the presenters were: What are short term opportunities and policy needs, and how do they relate to long term goals? What roadmaps, strategies, and other organizing frameworks can be used in thinking about the transition? What is the government role? Are there models for successful policy intervention of this magnitude and duration based on experience with other alternative transportation fuels promoted in the 1990s? What can be learned from public policies and initiatives that have changed automotive characteristics, such as fuel economy standards? How are international experiences relevant to U.S. hydrogen debates?

Attendees included representatives from leading federal and state government agencies; major global automakers and oil companies; smaller hydrogen fuel and technology companies; national research laboratories, institutes, and universities; and public interest, environmental, and energy policy groups. Although most participants came from the United States, a number of participants traveled to Asilomar from Europe and Asia as well.

The conference organizers from the DOE and ITS-Davis sought to stimulate discussion of key issues and plausible pathways toward a hydrogen transition, rather than seek consensus on a recommended course for the United States. More than 30 speakers made formal presentations, followed by group discussions. The chapters in this book are drawn from these lectures or from the discussions they engendered. They are organized into six groups, as follows:

- The Hydrogen End Game
- Strategies for a Hydrogen Transition
- Hydrogen Entrepreneurship
- Government Policies
- The Lessons of the 1990s
- Alternatives to Hydrogen Today

The remainder of this chapter summarizes the individual chapters from the Asilomar Conference. Biographical information about the author(s) of each chapter appears in Appendix A.

The Hydrogen End Game

Although the Asilomar conference was designed to discuss practical steps to start along a path toward a hydrogen transportation energy economy, several presenters jumped ahead and examined what a completed hydrogen transformation might look like. According to David Scott of the Institute for Integrated Energy Systems at the University of Victoria in Canada, the distant future is a good place to start a discussion about a transition to hydrogen because the long term goal is, in fact, much clearer than the pathways to get there. As he says, "I can't be sure if I'll catch a cold next month. But I'm sure I won't a hundred years from now."

Scott believes oil use as a transportation fuel will be outlawed by the middle of the century to save whatever resources remain for use in creating drugs, building materials, and even synthetic food. Coal use will also be banned for environmental reasons. Remaining primary energy resources, mostly renewable resources and nuclear power, will be harnessed to create either electricity or hydrogen energy currencies. Nearly all energy end uses late in the twenty-first century will be met by the intertwined electricity and hydrogen energy currencies. A key goal of the global transportation energy industry will be to convert electricity into hydrogen when energy storage is needed and to reconvert hydrogen back into electricity when a FCV needs to move.

Scott ultimately views the hydrogen transition primarily as completing the electric revolution begun more than a century ago. Because hydrogen fuel cells can be viable generators of power in motor vehicles, one of their key applications will be in transportation, but his palette is much larger.

Ken Kurani and his associates at ITS-Davis also paint a very large picture. They see hydrogen FCVs as far more than a successor to the conventional motor vehicle. To them, FCVs could form a system that fully integrates automobility, electricity, and information. This will be accomplished, in part, by the transformation of automobiles from their current design and role as primarily mobility tools. In a technological sense, automobiles will become integrated mobility/electricity/information platforms; in a behavioral sense, they will become mobile activity locales, new sites for a range of activities now conducted in homes or business offices.

This blurring of the distinction between transportation and other human activities will be one of the landmarks of an ultramodern world envisioned by Kurani and his colleagues. One of the behavioral and technological integrations is mobile electricity, the integration of electric drive, energy storage and delivery, and mobility technologies such that it is possible for the vehicle to deliver electricity for nonpropulsion uses, whether it is stationary or mobile. They argue that changes of this nature are needed to distinguish FCVs from their ICE competitor. If they compete solely on environmental benefits, the transition will be much longer and slower.

Strategies for a Hydrogen Transition

The first full function modern FCV was built in the early 1990s. The tenth FCV was built around 1996, and the hundredth FCV was built shortly after the turn of the century (USFCC, 2003). With a tenfold growth in FCVs occurring roughly every 5 years, there would be 1000 FCVs on roads well before the end of this decade, and 100,000 FCVs early next decade. The most aggressive scenario contemplated in a 2004 National Academies study of hydrogen is large scale production of FCVs beginning in 2015, increasing to 2 million FCVs on U.S. roads in 2028, and 10 million a decade later (NRC and NAE, 2004). Many industrial development experts suggest that annual production levels of 100,000 FCVs will be needed before major economies of scale come into play and production costs drop sharply. If the current pace of development continues, this might be achieved sometime between 2015 and 2020.

Even with rapid growth in hydrogen FCVs, they would not compete economically with gasoline and diesel fuels or with ICEs for a long time, and not even then unless new manufacturing systems are devised and fuel distribution systems are restructured. Each 2003 FCV cost upward of \$1 million to hand build. The infrastructure to refuel these vehicles consists of only a couple dozen costly compressed or liquefied hydrogen fueling stations.

With little prospect for short term profit, how can companies justify building hydrogen fueling stations and manufacturing FCVs? How can they justify sinking millions and even billions of dollars per year into something that will not generate substantial revenue, let alone profit, for many, many years? Three authors have given their perspectives on this question.

Toyota Motor Corporation is one of many major automakers that has built and is currently field testing a small handful of FCVs. As discussed by Taiyo Kawai, from Toyota's headquarters in Japan, the company will continue to accelerate the research and development of FCV technologies and to test this technology in a growing fleet of demonstration vehicles.

Where Toyota's business strategy differs from most others is its commercialization of vehicles with hybrid electric drivetrains, a key enabling technology for FCVs. By the end of 2003, Toyota had already sold more than 100,000 Prius hybrid electric vehicles (HEVs) in Japan and the United States, and had just launched a second generation model with much improved technology. The first generation Prius won the Society of Automotive Engineer's "Best Engineered Car of 1999" award and the second generation model won *Motor Trend* magazine's 2003 Car of the Year award.

Toyota intends to use most of the hybrid electric drive system developed for the Prius in its FCV power trains. The major difference is that the ICE in the hybrid drivetrain will be replaced with a fuel cell power unit. Toyota believes that by reducing the cost and improving the performance of hybrid electric drivetrain technology (what it calls Hybrid Synergy Drive) through commercial introduction in HEVs, it is accelerating the commercial advent of FCVs. Toyota is unique among automakers in proposing to introduce hybrid drivetrains to its entire lineup of cars and light trucks within a decade.

Joan Ogden from ITS-Davis takes a step back from automotive manufacturing to address the challenge of cost competitively distributing hydrogen fuel from production sites to vehicle refueling stations. She notes that hydrogen can be produced from a variety of feedstocks with zero or near-zero emissions—at scales suited to households up to entire cities.

In her view, a future hydrogen energy system will be interdependent with the rest of the energy system, especially the electricity system. To fully realize hydrogen's multiple benefits over the long term, it will be important to use hydrogen supply pathways with near-zero emissions of greenhouse gases and air pollutants, and to use diverse primary energy supplies. Ogden believes promising long-term options that could reach both low cost and zero or near-zero emissions include fossil hydrogen production with carbon dioxide (CO_2) sequestration, renewable hydrogen from biomass gasification or possibly wind powered electrolysis, and hydrogen from off-peak power based on carbon-free electricity.

Virtually all renewable and near-zero hydrogen supply options are more expensive than fossil hydrogen—with current production technology and current low volume distribution systems. Natural gas might be an acceptable transitional source of hydrogen over the next few decades because of its relatively low cost and low emissions. The use of hydrogen from natural gas in advanced hydrogen vehicles, versus using fossil derived liquid fuels in improved internal combustion engine vehicles, would provide rather small greenhouse gas benefits but would result in large reductions in oil use and tailpipe emissions of air pollutants. This chapter provides an understanding of the structure of a hydrogen system, how it might evolve, and its benefits and costs.

Although coal did not rank high on the list of preferred hydrogen sources of some presenters at Asilomar, Richard Doctor from Argonne National Laboratory and his colleague John Molberg believe this abundant fossil fuel can be tapped cleanly and efficiently for hydrogen. Based on a detailed analysis of a coal-to-hydrogen production facility now in operation in Beulah, North Dakota, they conclude that the economics of coal to hydrogen look favorable if the 2003 prices for natural gas remain the norm. They also conclude that estimated hydrogen costs will be $1.30 to $1.40 per kilogram (one kilogram has about the same energy as one gallon of gasoline), and that CO_2 can be sequestered. The greatest challenges will be the transport of CO_2 and the availability of CO_2 sequestration reservoirs. Here the regulatory landscape is still not fully formed.

Hydrogen Entrepreneurship

Although their work is frequently overshadowed by announcements from major automakers and oil companies about hydrogen R&D projects, there exists a small industry in the United States today that produces and sells hydrogen, fuel cells, and FCV components. This industry struggles to sell precommercial fuels and technologies in today's commercial markets. With no conventional line of business to fall back on to absorb R&D costs, the hydrogen and fuel cell industry must find customers today to keep their business afloat.

Always a high risk industry, hydrogen companies have become even harder to keep afloat since a U.S. recession beginning in 2000 seriously limited access to venture capital and other sources of investment capital. Two chapters discuss hydrogen entrepreneurship in hard times, one by a leader of a company in the field and one from an academic specializing in industrial innovation.

Chip Schroeder is CEO of Proton Energy Systems, Inc., a Connecticut-based start-up company that sells small scale hydrogen production units that use a unique electrolysis technology to split hydrogen from water. The challenge he faces daily is to find customers that buy small quantities of hydrogen and then convince them that his on-site hydrogen generator is a better way to obtain the hydrogen than the current practice of buying storage containers filled with hydrogen delivered by truck from one of the scattered industrial gas manufacturing sites in the United States.

This challenge can be met, says Schroeder, when an economic case can be made for on-site electrolytic hydrogen production. For example, if the electricity used to make electrolytic hydrogen comes from low priced, off-peak coal or nuclear power production, and if the hydrogen is then used to replace high priced fuels such as gasoline, these coal or nuclear resources have effectively been converted into transport fuel. In such practical applications, the economic value added can offset the net energy loss of the process. An even more compelling justification for electrolysis comes from the desire to see renewable power make an impact on transportation markets. Renewable resources, such as wind, solar, or geothermal power, provide electricity but not automotive fuel. The only practical way to turn renewably generated electricity into fuel is through electrolysis.

Another key to competitive electrolytic hydrogen is on-site production, which minimizes or eliminates costly truck transport costs for delivered hydrogen. In effect, electrolysis takes advantage of the existing infrastructures for electricity and water. The cost of an electrolyzer sited at a vehicle fueling station and sized to fill 10 to 20 cars per day is far less than the total capital cost of a new large scale steam methane reformer that requires a new pipeline or truck-based delivery infrastructure. For this reason, on-site electrolysis could have a role in the introductory stages of the hydrogen fueling marketplace. The longer term role of electrolysis for fueling will depend upon how the economics of converting electricity to hydrogen compares with the economics of other fueling options.

David Bodde, a professor at the University of Missouri and author of a new book, *The Intentional Entrepreneur: Building the Real New Economy*, stands back from Schroeder's daily business struggles and analyzes the dynamics of new market creation. He believes the nurturing of entrepreneurial activity and private investment in hydrogen businesses could be vital to a successful transition to a hydrogen transportation energy economy. He notes that entrepreneurs follow opportunity, usually seeking to establish an economically significant business within about 10 years. This allows investors an opportunity to recover their cash from the enterprise, either through sale to a larger firm or a public stock offering.

Bodde believes that public policy is difficult to tailor to the strands of opportunities entrepreneurs follow. Perhaps the strongest policy need, therefore, is constancy of intent—a set of public priorities that are likely to stay in place for the 10-year planning horizon of the entrepreneurs and their investors. Entrepreneurs can then direct their investments accordingly. Other effective policies would seek to align public needs with private markets, allowing the societal preferences for clean air and fuel security to be expressed in customer preferences as well.

A mandatory cap on carbon emissions, for example, could be enforced by a requirement that suppliers and users of fossil fuels hold tradable rights for each ton of carbon they produced—the so-called cap and trade approach. This would have the effect of placing a price on carbon emissions that would eventually be included in the price of all goods and services. Environmentally benign technologies, like those based on hydrogen, would thus gain an advantage over those that are less benign. Taxes on the burning of specified fossil fuels would have the same effect, he notes.

Government Policies

The issues raised by a fundamental transformation of the world's transportation energy economy clearly go well beyond hydrogen and FCVs. Other issues include threats to global energy security from oil dependence, the protection of public health from the pollutants in automotive exhausts, and the integrity of the planetary ecosystem and social systems as climate changes. Nearly every speaker at Asilomar highlighted the appropriateness of government actions to affect the pace and direction of a hydrogen transition. Three chapters in this book discuss the role of government. They are written by leaders of government agencies that have major hydrogen energy commitments: the DOE representing U.S. government programs, the California Energy Commission, and a European agency in The Netherlands that is involved in hydrogen programs throughout the European Union (EU).

According to Stephen Chalk, the DOE Program Manager for Hydrogen, Fuel Cells and Infrastructure Technologies Program, and co-author Lauren Inouye from Sentech Inc., the Hydrogen Initiative and FreedomCAR programs announced by President Bush early in 2003 plan to invest $1.7 billion in federal funds over five years to develop a more sustainable energy and transport system. Chalk and Inouye believe that hydrogen FCVs are a promising technology, but they are only one element in the larger U.S. strategy to secure a healthy environment and sound economy for future generations. It is important, they emphasize, that hydrogen not be "oversold" to the public, leading to unrealistic expectations about either the time frame of hydrogen FCVs or the scope of problems they can help address.

The current DOE hydrogen and fuel cell strategy includes major investments in research, development, and demonstration programs over the next decade. The current program plan, however, does not even call for a DOE decision about the viability of using hydrogen as a transportation fuel before 2015. According to the authors, wide adoption of FCVs will most likely not take place until 2020, and a minimum of 20 years will be required for the majority of petroleum-based ICEs to be effectively replaced. The DOE time frame and budget allow the agency to support the exploration of multiple production, delivery, and storage pathways, and to reevaluate the feasibility of different options with phased "go/no-go" decision points.

Since the establishment of the zero emission vehicle (ZEV) program by the California Air Resources Board (CARB) in 1990, California has pressed industry to develop and commercialize advanced vehicles, including hydrogen FCVs. Jim Boyd, a commissioner at the California Energy Commission (CEC) and former executive director of CARB, discusses the California program's positive effects on a hydrogen transition and summarizes two new CEC and CARB reports on the state's growing oil dependence. These analyses show that by far the most effective strategy to reduce petroleum consumption is increased new vehicle fuel efficiency. A secondary strategy, however, involves the substitution of hydrogen FCVs and other alternative fuel or advanced vehicles to account for up to 20 percent of new light duty vehicle sales in the state. These two strategies implemented together could yield significant reductions in motor vehicle emissions and fuel use in the 2020 to 2030 time frame.

Although the California transportation programs have been controversial and frequently drawn opposition, Boyd notes that the biggest energy and automotive companies remain at the table in California with government agencies, infrastructure companies, and other stakeholders talking about hydrogen, fuel cell vehicles, and all the programmatic needs that must be addressed to complete the road to a hydrogen economy. At no other time in the state's history have the energy and auto companies joined to facilitate a change to an alternative-fuel-powered advanced-vehicle technology.

Among those listening to the lectures on U.S. federal government and California state hydrogen initiatives at Asilomar was Barend van Engelenburg from the Ministry of Environment in The Netherlands. He provided a chapter summarizing the European effort to promote a hydrogen transition. His chapter says that focused attention to hydrogen among countries in the European Union started in 2000, considerably later than in the United States, with the publication of a Green Paper on energy security, which identified hydrogen as a potentially important transportation fuel. A later White Paper on transport policy emphasized the intent of the European Union to deal with alternative fuels, especially hydrogen. The European Union has a common and coordinated R&D program, called the Framework Program (FP), which is now dispersing about €144 million (Euros) to fund fuel cell or hydrogen projects. In the course of 2002, the European Commission (EC) working on behalf of EU countries appointed two groups that are now studying hydrogen transition issues. In 2003, a conference on hydrogen was held in which EC officials indicated a firm commitment to hydrogen. A collaboration agreement between the EC and the U.S. DOE was signed in November 2003.

The hydrogen movement in the European Union is similar to that in the United States in that it represents an economic and political collaboration of states. The differences, however, are much more pronounced. Individual European states are much more independent than states in the United States, and the EU parliament has much less formal power than the U.S. Congress. Each country is free to enact the provisions of EU directives in its own way according to national rules and habits.

Overall, van Engelenburg believes the European Union has a much weaker commitment to hydrogen than the U.S. government. It is pursuing a program that relies more heavily on a consultation process with select industry leaders, while the United States has chosen a structured approach, proceeding stepwise from a basic resolution to a vision, roadmap, implementation plans, and external review. The structure of the process in the European Union is not yet clear. There is no basic formal resolution or decision about hydrogen within European governments, he says.

The Lessons of the 1990s

The Alternative Motor Fuel Act of 1988 was the first major federal law in the United States to address alternative transportation fuel use. It was followed by the Clean Air Amendments Act of 1990 and the Energy Policy Act of 1992, which greatly expanded alternative fuel use mandates and fuel quality specifications. Over the past decade, transportation fuel and vehicle providers have struggled to comply with a bewildering array of federal government requirements. Despite some considerable successes and the presence of over 400,000 alternative fuel vehicles (AFVs) on U.S. roads, their population remains at less than one half of one percent of the total motor vehicle fleet. Thus, the country remains nearly as far today from diversifying its transportation fuel and vehicle mix as it was 15 years ago.

The experiences of the past few decades, however, provide a rich and unique historical perspective to examine new government transportation strategies. Three Asilomar presentations look to past experiences for lessons learned. The first two chapters are by alternative fuel veterans from the DOE and DaimlerChrysler, and the third by two experienced alternative fuel researchers. All three chapters assess past experiences with technology development, investment, and regulation and policy. They offer perspectives on how hydrogen and fuel cell technologies can best be promoted by the government to avoid the pitfalls that hampered past efforts.

David Rodgers and Barry McNutt, senior policy analysts at the DOE, provide a more qualitative assessment of what happened. They note that alternative fuel advocates made a convincing case in the late 1980s and early 1990s that alternative fuels and vehicles could reduce emissions and oil imports. It was not to be. When AFVs were mandated, the automotive and oil industries, with significant sunk investments in conventional fuels and vehicles, quickly responded. Oil companies reformulated gasoline to reduce emissions across the entire vehicle fleet, and car companies continued to reduce emissions from conventional vehicles. They conclude that, while AFV mandates did not directly reduce emissions, they did so indirectly by motivating improvements in gasoline and vehicle technology.

Rodgers and McNutt also address market niche strategies. They observe that vehicle fleets, originally considered one of the best early niche markets for alternative fuels, proved resistant. They conclude that, even if successful in fleet applications, the broader record of the 1990s suggests that AFV use in niche markets doesn't necessarily grow into a mainstream market for light duty vehicles, even with significant government support. And thus, they suggest, hydrogen and fuel cell vehicles will need an early and significant presence in mainstream consumer markets to serve as a basis for a successful transition. To achieve this, policies that value the social benefits of hydrogen and FCVs will be required to stimulate consumer demand enough to achieve the necessary timetable and scale of transition.

Bernard Robertson and Loren Beard, senior executives at DaimlerChrysler, provide a different perspective. Chrysler Corporation, prior to its merger with Daimler Benz, sold a wide range of AFVs in the U.S. market, including cars fueled by natural gas, propane, methanol and ethanol, and a battery powered minivan. The authors conclude, based on this experience, that any government sponsored AFV program needs to be an integral part of a clearly articulated and broadly accepted national energy policy with explicit goals. The goals should delineate with some precision the problem to be tackled—whether it is CO_2 emissions, fossil fuel use, or dependence on imported petroleum. The goals and targets of the transportation sector must be set at achievable and measurable levels, and must complement goals and targets for other segments of the economy, so that the most cost effective, least disruptive solutions can be pursued.

Most importantly to Robertson and Beard, any AFV program must compete. Every major automaker must be willing to subsidize the development and early deployment phase of new technology introduction, but fiscal responsibility requires that there be some promise of a positive return in the future. Since most AFVs are not economic today, government must recognize that it needs to play a role in establishing economic policies to "level the playing field" against widely available, inexpensive petroleum fuels and the very mature internal combustion engine. This may include providing new financial incentives or subsidizing both vehicles and fuels. As important as these incentives and subsidies are, they must be finite in duration. Robertson and Beard emphasize that any replacement for the gasoline-powered spark-ignited engine must eventually be cost competitive if it is to be successful.

Finally, Jonathan Rubin from the University of Maine and Paul Leiby from Oak Ridge National Laboratory collaborated on an economic assessment of the effect of several government financial incentive programs to encourage the purchase of AFVs and HEVs. In analyzing the transition to alternative fuels other than hydrogen—ethanol, methanol, natural gas, propane, and electricity—their economic model led to some important conclusions that bear on the proposed hydrogen transition. For AFVs, the most important barriers seem to be limited fuel availability and vehicle scale economies. For HEVs, incremental vehicle costs are large. As a result, vehicle scale economies matter, but scale cost reductions are more easily attained by the use of widely shared components—such as batteries, motors, and controllers—across multiple vehicle platforms. Similar gains should be possible for FCVs.

Rubin and Leiby concluded that transitional barriers will prevent any significant alternative fuel use by 2010 unless there are sustained, expensive market interventions by the government. These market barriers are approximately equivalent to a cost of $1.00 per gallon today and will persist at a level of $0.50 per gallon in 2010. These transitional barriers are likely to have similar implications for prospects of hydrogen ICE vehicles and for FCVs.

Alternatives to Hydrogen Today

Even the most ardent supporters of a hydrogen transportation energy economy admit that it will be at least a decade and probably longer before there are enough hydrogen FCVs on the roads to make a serious dent in oil use or to improve urban air quality. Some people, including two presenters at Asilomar, believe that other transportation technologies can be implemented much more quickly, providing much more progress toward the goals of oil displacement and environmental quality.

John M. DeCicco, a transportation policy analyst at Environmental Defense, suggests in Chapter 15 that the paradigm shift to hydrogen rests on a set of beliefs that are not yet borne out. He argues that it is uncertain when, or even if, hydrogen and fuel cell technology will be competitive, suggesting that a focus on hydrogen is premature at best. DeCicco also raises a deeper question about whether a technological response to transportation energy problems can really be successful without a widely held public commitment to fix the problems the new technology addresses.

In DeCicco's opinion, a public commitment to replace oil as a transportation fuel has not yet been made on either energy or environmental grounds. Political commitment to cut oil use or greenhouse gas emissions seems essential before any new technological solution can be brought into being, no matter how promising it may be on paper. Absent these commitments, DeCicco argues a more practical strategy may be to focus on implementing widespread fuel economy improvements and emission reductions in conventional vehicles.

Paul MacCready, founder and head of Aerovironment in California, is considered by many to be the "father" of the General Motors EV-1. His company has built and operated many of the world's most innovative electric vehicles, including both battery and fuel cell powered ground and air vehicles. Having watched the rise of battery electric vehicle technology in the 1990s and their subsequent decline in favor of FCV technology this decade, he questions whether the transition to hydrogen might be occurring a little too quickly.

MacCready believes that recent advances in lithium batteries warrant another look at the potential of battery electric vehicles. Some lithium batteries now store nearly 200 Watt hours per kilogram (Wh/kg) of energy and have very high efficiency levels during charging and discharging cycles. Additionally, with a bidirectional charge–discharge hookup, the batteries can be used to help an electric utility adjust its grid capacity and handle peak loads.

A reasonable initial deployment strategy, says MacCready, should begin with an HEV with an electric battery power range of 75 to 100 miles. Use of battery electricity obtained from the grid through recharging should be able to handle about 90 percent of the annual driving distance of such a car, while using gasoline or an alternative fuel to augment power during longer trips of up to 400 miles. As battery costs decline and performance improves, the battery range can be extended and the ICE engine emissions decreased or eliminated.

Summary

This book addresses an extraordinarily complex and far-reaching issue facing the world today. It is not just a question of which fuel to put in our tanks. It is much more. It is about transforming two of the largest industries in the world—transforming everything from upstream extraction and manufacturing, to downstream retailing and financing. It is about integrating stationary and mobile energy, in effect merging electricity and hydrogen into one system. And it is even more than that.

The transition to hydrogen involves transforming our society—economically, socially, and environmentally, locally as well as globally. It involves choices that affect almost everything—not only energy security and environmental quality, but also economic growth and social well-being. The question of hydrogen is not a narrow, technical energy discussion.

The chapter authors all agree the future is uncertain, and that policy initiatives are necessary and appropriate to address transportation energy concerns. All seek durable and consistent public policy, and none opposes a long term hydrogen transition strategy. That is where the agreement ends. Some want more attention today to hydrogen, others less. Some urge immediate public support for initial hydrogen and fuel cell investments; others counsel patience and delay.

The debate over hydrogen is just beginning. It will be a long and continuing debate. This book provides an initial foundation and framework for that debate. It provides state-of-the-art knowledge of what we know about fuel cell vehicle technology, hydrogen supply options, and policy. But mostly what it provides is informed insights by experts and leaders from a broad range of beliefs, interests, and values. And therein lies the chief benefit of this book. It is a primer on what is known and believed, and an assessment of what is possible and desirable.

References

Cannon, J. 1995. *Harnessing Hydrogen: The Key to Sustainable Transportation.* New York, NY: Inform, Inc.

Davis, S., and S. Diegel. 2003. *Transportation Energy Data Book.* Oak Ridge, TN: Oak Ridge National Laboratory.

National Research Council and National Academy of Engineering. 2004. *The Hydrogen Economy: Opportunities, Costs, Barriers, and R&D Needs.* Washington, DC: National Academy Press. (Available online at www.nap.edu)

U.S. Central Intelligence Agency. 2001. *Global Trends: 2015.* Washington, DC: U.S. Central Intelligence Agency.

U.S. Energy Information Administration. 2003a. *Annual Energy Review.* Washington, DC: U.S. Department of Energy.

U.S. Energy Information Administration. 2003b. *International Energy Annual.* Washington, DC: U.S. Department of Energy.

U.S. Environmental Protection Agency. 2003. *Latest Findings on National Air Quality: 2002 Status and Trends.* Washington, DC: U.S. Environmental Protection Agency.

U.S. Fuel Cell Council. 2003. Website fuel cell vehicle directory, http://www.usfcc.com.

CHAPTER 2

Back from the Future: To Build Strategies Taking Us to a Hydrogen Age

David Sanborn Scott

I can't be sure if I'll catch a cold next month. But I'm sure I won't a hundred years from now. Sometimes, for some things, we can project the distant future better than tomorrow—an observation particularly relevant for our energy system, where looking further out parts the fog of immediacy and allows us to leap the tangles of today's conventional wisdom and "wishdom," helps provide clearheaded strategies for what we should be doing today. That is the theme of this lead chapter in a book that analyzes a global transformation to a hydrogen-based energy system.

To take such a long view, we'll first examine the question, "why hydrogen?" To provide context, we'll then introduce a five-link energy-system architecture and follow this with a discussion of a chain that forms the backbone of any energy system's architecture. Emphasis will be placed on the role of energy currencies, the central link in these systems. Energy currencies are energy forms that are not naturally occurring. They are produced from a variety of natural resources and can be applied to a wide range of energy applications. Electricity is the great energy carrier of the twentieth century. Hydrogen will emerge as the great energy carrier, alongside electricity, in the twenty-first century. The chapter will close with a description of the sources, infrastructures, and service technologies likely to characterize the deeper energy future that will be built in the twenty-first century and serve human needs for centuries to come. Based on this long view, I propose a number of near-term strategies to guide energy policy decisions.

Why Hydrogen?

As we embark upon the twenty-first century, two specters threaten global civilization. One is the prospect of economic disarray and incitement to war caused by the local depletion and global maldistribution of high quality fossil fuels, especially oil. The second is the prospect of almost unimaginable environmental, economic, and cultural disruption caused by climate volatility and triggered, primarily, by our energy system's carbon dioxide effluent.

If we remain tethered to fossil fuels, it is merely a question of when, not if, one or both specters become reality. Today, we're seeing the precursors of both. From the viewpoint of Earth's physiology, climate volatility is much more critical. This suggests that the global environment, not energy resource depletion, is the absolute cap on fossil fuel use. Nevertheless, at any moment, flaring geopolitics could suddenly annihilate the global fossil energy supply system.

Happily, a straightforward pathway can steer us clear of both specters. We must rapidly adopt sustainable energy sources that don't emit carbon dioxide. Such sources include solar, wind, and safe, economic next-generation nuclear power. Perhaps more critical—since it's less well understood—we must rapidly adopt the twin energy currencies, hydrogen and electricity. Alone among currencies, both can be manufactured from any energy source, neither emits carbon dioxide, and together they can provide the full menu of civilization's energy services, from fueling airplanes to running computers.

These few paragraphs encapsulate both the global energy problem and its solution. Yet before we move on to the main themes of this chapter, we should explore, just a little deeper, "why hydrogen?" Today's transportation fuels are harvested exclusively from fossil energy sources. Because fossil sources are threatened by regional depletion and are a sure cause of international conflict, civilization must move to sustainable, regionally available, nonfossil sources. This means we must be able to harvest sustainable sources so they can manufacture chemical fuels in general and transportation fuels in particular. Realistically, the only way to manufacture sustainable sources of chemical fuels is via hydrogen. How else, for example, can we use the energy from wind, solar, or nuclear sources to fuel an airplane? Finally, a significant move from fossil to sustainable sources can only begin with the increased use of hydrogen in transportation and can only be completed with the supremacy of hydrogen among chemical fuels.

Next consider the issue of climate volatility, where a parallel logic again takes us to hydrogen. The carbon dioxide effluent from fossil fuel consumption is pushing our planet toward climate destabilization that, if unabated, will be catastrophic. To eliminate civilization's carbon dioxide emissions, we need both noncarbon emitting energy sources and noncarbon energy currencies. Many noncarbon sources are available or can be developed, including hydraulic power, sunlight, wind, nuclear fission, and perhaps fusion. By contrast, there are only two noncarbon currencies that can together supply the full menu of civilization's energy services. These currencies are electricity and hydrogen. Electricity is already established—but of course airplanes can't use electricity as fuel. Therefore, anthropogenic carbon dioxide emissions can only be slowed by the extensive use of hydrogen and can only be eliminated with the supremacy of sustainable hydrogen over other chemical fuels.

Hydrogen and Electricity

A little over a century ago, electricity entered the marketplace as a new energy currency. Not an energy source itself, electricity is manufactured from an energy source, and the resulting energy is then transferred to be used in an enormously versatile range of applications. Electricity replaced the direct use of power from a steam engine, the major energy provider of the nineteenth century. The twentieth century saw the increasing electrification of energy processes in our homes, businesses, and industries. Free range transportation is the only major energy sector that has not yet yielded to electricity.

Hydrogen is now emerging as a kindred spirit to electricity. It is a second energy carrier analogous to electricity, sharing many synergies with electricity. Like financial currencies, for example, you can exchange energy currencies by converting electricity into hydrogen and vice versa. There are many other synergies between hydrogen and electricity that, in turn, allow the two currencies to provide the full menu of services civilization demands. The patterns in these synergies are:

- Hydrogen can be a transportation fuel or material feedstock. Electricity cannot.
- Electricity can be used to transmit, process, and store information. Hydrogen cannot.
- Hydrogen can be stored in enormous quantities. Electricity cannot.
- Electricity can transport energy without transporting material. Hydrogen cannot.
- On Earth, hydrogen will be best for long distance energy transport. In space, electromagnetic radiation, as a proxy for electricity, wins.

The fact that hydrogen can be stored in enormous quantities, the third point, confers its future as both the staple transportation fuel and the system's energy storage sponge. Storage is needed to improve the reliability and efficiency of our energy system. It is needed to allow production of energy currencies when resources are available, regardless of demand, and to hold these currencies until needed. Electricity cannot provide this; hydrogen can. This is especially important for renewable sources—like when we harvest solar during the day but need its energy at night. And of course, we will need to store energy for national security. Electricity cannot fill the storage role.

Turning to the fourth point, environmental intrusion is always carried in material. So it's interesting that to deliver energy via electricity we don't move material. Although transfer of hydrogen energy requires material movement, such as the flow of hydrogen gas through a pipeline, hydrogen carries more energy per unit mass than does any other common currency.

These hydrogen–electricity synergies can help us predict how each currency will capture different markets. Hydrogen will fuel free range surface vehicles, while electricity will "fuel" computers. We can now pick any imagined task and, by applying our template, anticipate whether hydrogen or electricity is likely to win—or if both will have a shot.

Now let's consider how hydrogen and electricity are "kindred spirits" with shared characteristics. At least these three are important:

- Both hydrogen and electricity can be manufactured from any energy source.
- Both hydrogen and electricity are interconvertible. (Electricity can be made from hydrogen, hydrogen from electricity.)
- Both hydrogen and electricity are renewable.

Most of us know that electricity can be generated from any energy source. Hydrogen can also be harvested from any source—demonstrated, if you like, because hydrogen can be generated by electricity via electrolysis. Of course, there are many ways to harvest hydrogen without using electrolysis: for example, by steam reforming of hydrocarbons such as natural gas, the most common route today.

The second similarity between hydrogen and electricity is that both are interconvertible. Electrolyzers convert electricity to hydrogen. Fuel cells convert hydrogen to electricity. Other fuels are not mutually interchangeable with electricity. Gasoline, for example, can be converted to electricity. Alternators on board motor vehicles do this every day. But electricity cannot be converted to gasoline. The road connecting electricity and gasoline only allows one-way traffic. It creates a dangerous, systemic brittleness in our transportation energy system.

The third way hydrogen and electricity are similar speaks to our hankering for renewables. People sometimes say to me, "The good thing about hydrogen is we have so much water that we'll never run out of water to make hydrogen." That misses the point. It's like saying, "The good thing about electricity is we'll never run out of electrons." Electricity is generated by separating electrical charges, which later, when the electricity is used, come back together to give charge neutrality. Analogously, hydrogen is generated by splitting water into its components, hydrogen and oxygen, which later, when the hydrogen is used, combines with oxygen, returning the water.

Architecture of the Energy System

Most energy experts now agree that sometime during the early to middle decades of the twenty-second century, and perhaps much earlier, sustainable sources transformed into hydrogen and electric energy currencies will dominate civilization's energy system. To better appreciate this future, we will find it useful to have a concept of our energy system's architecture to identify those links we can reasonably predict and those we'd be fools to even try.

The strange thing about the energy system is that we almost never think about it as a system. Instead we think about its bits and pieces—about electricity networks or oil cartels, airplanes or fax machines—but not about the system. To understand how our energy system works, we need a picture of its architecture, how the bits and pieces fit together, and what they do for each other.

Let's look for an architecture that is as simple as possible. Consider a five-link chain that starts with energy services, moves to technologies that deliver the services, then to energy currencies that feed the service technologies, then transformer technologies that produce the currencies, and finally the sources that the transformer technologies harvest. Such a chain is shown in Fig. 2-1.

Unfortunately, many people view our energy system as simply the three rightmost links, which is the energy "sector" report in the financial pages of your local newspaper. So there is information about oil production or electric generation, but little attention to people's changing desires for energy services. This view truncates the leftmost links in the systemic architecture. We shouldn't forget that people want energy services, not prescribed levels of oil production. They want to be able to cook a meal, watch the news, drive to grandma's house, or keep warm. Energy is a means to these ends. Services are the ends. It isn't that people want the delivery of these services to consume large quantities of energy; indeed, they're often pleased if the services require less energy.

Most energy analysts tend to plod through the energy architecture from sources to services because that is how the energy flows. But energy flow is counter to the way the system is driven. The process begins when people demand a service, which demands a service technology, then an energy currency, and so on back to energy sources.

In the transition to hydrogen, it's the demand for improved energy services that will drive the system, not the availability of energy sources.

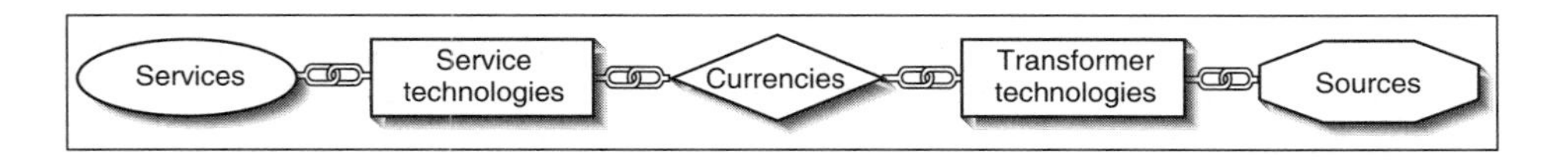

FIGURE 2-1. Energy System Architecture.

The world has been using less energy per unit service every decade since the Industrial Revolution. This trend will continue sweeping us into the future as more service will continue to be provided for less energy.

In order for hydrogen systems to become viable, five criteria for sustainability must be met throughout each stage in the energy architecture. The idea of sustainability has been described in *Our Common Future*, a report of the World Commission on Environment and Development, as a system that meets "... the needs of the present without compromising the ability of future generations to meet their own needs." I have always been uneasy about the word "needs" because it requires an answer to the ancillary question: by whose criteria? Moreover, it seems to me we should aspire to something more than mere existence, which the word "needs" implies. So I have modified this definition with a few words of my own to create an alternate definition: Sustainable development improves lifestyles—by bringing cultural and economic growth embedded within environmental gentility—without jeopardizing the ability of future generations to live even better.

Next, we must address how the definition can be used as a foundation for criteria explicitly designed for energy systems. I think first rank criteria come down to evaluating five aspects of the hydrogen energy system architecture. First, regarding the quality of the energy service, hydrogen service initially cannot be inferior to an equivalent service provided by established systems, and it soon must be significantly better. Second, the environmental gentility of the system chain must ensure that inputs and outputs to and from each link of the system chain minimally intrude upon nature's flows and equilibriums. Decommissioning, or returning industrial land to green space and recycling used materials, must be both technically and economically feasible. Third, the economic cost of the service may, for a short time, be somewhat greater than the cost from established, competing systems, but soon the cost of services must be lower. Fourth, the energy source used to produce the hydrogen must be available into the foreseeable future, with global distribution relatively uniform so as not to aggravate geopolitical tensions that risk wars or economic blackmail. Finally, as much as possible, the hydrogen system must be resilient to geopolitical, technological, economic, and environmental surprises.

Let's examine the hydrogen economy with respect to just one of these criteria—environmental intrusion—to see how hydrogen does. While the need to include the complete system should be obvious, there are nuances. First, environmental intrusion may be exported from one link to other links, either up-system or down-system. Consider a hydrogen-fueled city bus using a fuel cell power train. The fuel cell bus will improve urban air quality because it releases only small amounts of clean water vapor, rather than soot, nitrous oxides, and carcinogens, all of which spew from its predecessor, a diesel bus. But what if the hydrogen was produced by electrolysis using electricity generated from a coal fired station? If the generating station was far from the city, the urban environment will still benefit, but the air quality near the generating station will be degraded. Moreover, carbon dioxide will be dumped into the global commons. By manufacturing the hydrogen from coal, we push the environmental intrusion downsystem—push it from the service technology, such as buses in the city, to the transformer technology, in this case hydrogen production from coal on the outskirts.

One way or another, environmental intrusion always involves material that intrudes upon nature's flows or equilibrium—material taken from the environment, diverted within the environment, or put into the environment. Knowing environmental intrusion is rooted in material, we set out four questions:

- What is the magnitude of the intruding material compared with its natural level?
- What fraction of the global commons does the intruding material occupy?
- What is the characteristic residence time of the intruding material?
- How metastable is the disturbed equilibrium?

Regarding the first question, at the beginning of the Industrial Revolution carbon dioxide was present in the atmosphere at concentrations of about 270 parts per million by volume (ppmv). By 2000, carbon dioxide levels had risen to more than 370 ppmv. This is a change of more than 33 percent above preindustrial levels. All reputable studies have shown that, even with aggressive mitigation policies, carbon dioxide atmospheric content will have increased to 200 percent of its preindustrial levels by the turn of the next century.

Regarding the second question, anthropogenic carbon dioxide is dispersed throughout Earth's entire atmosphere and therefore permeates the Earth's entire biosphere. In answer to the third question, carbon dioxide is likely to remain about 200 years in the lower atmosphere, increasing to 400 years or more when it enters the upper atmosphere. Finally, the disturbed equilibrium in climate caused by adding carbon dioxide to the atmosphere is, indeed, fundamentally metastable.

Greenland ice core data allow us to track the global temperature and the atmospheric concentration of carbon dioxide and methane for some 420,000 years. During this period the planet has gone through periods warmer than today and periods much colder—when temperature and concentrations of these chemicals tracked each other like three dogs on a short leash. During these more than 240,000 years, the highest previous atmospheric carbon dioxide concentration was 310 ppmv. Since atmospheric carbon dioxide growth is overwhelmingly the result of the current fossil fuel energy system, the requirement to reduce and ultimately eliminate carbon dioxide emissions will become the environmental issue that dominates energy system development.

Hydrogen in the Distant Future

Just as it would have been impossible to anticipate cell phones and laptops at the turn of the twentieth century, today only a jester would attempt to predict the hydrogen consumer products of the twenty-second century. In contrast, I think we can get a good feel for the nature of exergy infrastructures, dominant energy sources, energy distribution and delivery systems, and the relative market share of hydrogen and electricity. And while unable to anticipate specific widgets, we should be able to anticipate some features of broad service "domains" like transportation—or where fuel cells, our "chip of the future," will be found, and where they won't.

Free Range Transportation

The reasons to focus on hydrogen applications in transportation are compelling. First, hydrogen will absolutely dominate electricity as the currency of choice in transportation because it can be stored onboard and replenished quickly during refueling. Second, in terms of quantities, transportation will always be a major energy user, and perhaps the major energy use sector of our economy.

In the distant future, I believe hydrogen will fuel all free range vehicles. Free range surface (land and sea) vehicles will be powered predominately by fuel cells. Aircraft powerplants will remain heat engines but will be fueled by hydrogen. Fixed route transportation systems, including ground-based mass transportation, will be a mixture of energy currencies. Hydrogen fuel cells will power free range buses, but subways and high density rail systems will likely continue to be powered by electricity.

Liquefied hydrogen is likely to provide onboard hydrogen storage in the distant transportation future. Liquefaction efficiency will be much improved, perhaps using magnetocaloric and other advanced technologies. Liquefied hydrogen has far less weight per unit of energy than any form of chemical fuel. It will be essential for aircraft, and simply better than other forms of stored hydrogen for surface vehicles.

The fact that liquefied hydrogen contains both thermomechanical and chemical "exergy" will be the key to important technical synergies. Energy is the theoretical maximum work that can be delivered when bringing an energy commodity—a source or currency—into equilibrium with its environment. The adjectives "thermomechanical" and "chemical" identify in what way the commodity is out of equilibrium with its environment. One example is aircraft, where the exergy-of-cold can be used to refrigerate aerodynamic surfaces, which Lockheed studies have projected will reduce fuel consumption by 30 percent and direct operating costs by 20 percent. To highlight synergies in dual-exergy content currencies in the case of aircraft, the thermomechanical exergy-of-cold, which constitutes 11 percent of exergy in liquefied hydrogen, provides three times as much "drive" per unit exergy than does the chemical exergy.

Cryogen exergy recovery systems (CERS) will become *de rigueur*. Various CERS technologies will be employed with fuel cell vehicles. Perhaps the simplest will use air separation for oxygen extraction to give the fuel cell an added acceleration punch. New materials and working fluids will permit "heat engines" to operate by taking heat from the environment and rejecting heat to the cryogen. Sometimes it may be feasible for a CERS to draw its heat from the liquid warm water effluent from fuel cells rather than from the environment. I've called such technologies enhanced cryogen exergy recovery systems (ECERS).

When using liquefied hydrogen, the ratio of thermomechanical to chemical exergy remains constant as the fuel is consumed. This allows CERS and ECERS to operate with constant thermochemical exergy ratios throughout the full refuelling cycle. In contrast, if compressed hydrogen storage is used, the ratio of thermomechanical–chemical exergy changes over the refuelling cycle—from a maximum when first refuelled to zero when the tank is almost empty.

Continental Energy Distribution

Today, the staple modes of transporting bulk energy across continents are high-voltage alternating-current (hv-ac) electricity networks, oil or natural gas pipelines, and railroads (in the case of coal). By contrast, during the fully developed hydrogen age, the primary continental energy transportation will be via gaseous hydrogen pipelines. Hydrogen pipelines will transmit energy with much lower energy losses than hv-ac can, and they are even more efficient than more expensive, higher efficiency, high-voltage direct-current (hv-dc) power lines. For distribution–delivery networks, gaseous hydrogen is more attractive than liquefied hydrogen for pipelines, not only because the capital cost of a cryopipeline is very high, but because liquefied hydrogen is an incompressible fluid and, hence, lacks the energy sponge properties of gaseous pipelines.

Inherently, hydrogen pipelines are wonderful energy storage devices. Therefore, they can both store the product of sources when the demand falls below supply and give up energy when demand rises above supply. Thus, they inherently load-level for constant output sources and can always absorb energy from intermittent and unpredictable sources.

In urban settings, we can expect pipes will deliver gaseous hydrogen directly to homes, offices, factories—and sometimes community "nodes." The electricity required for electric appliances, like computers, will normally be produced on-site from fuel cells. This will be direct current electricity—alternating current will essentially vanish because its prime value in today's world is for stepping voltage up and down as the electricity moves from trunk distribution to serving local loads. The single other advantage to alternating current is its inherent ability to keep electric clocks on time, but time accuracy can easily be done by many other means such as crystals.

Liquefied hydrogen taken from the gaseous hydrogen network will be liquefied at high capacity plants, perhaps located adjacent to nuclear power plants and then distributed by cryogenic tanker trucks or rail tankers to vehicle refueling stations. This scenario might be modified if there were significant advances in hydrogen liquefaction processes that improved efficiencies, lowered capital and operation costs, and were not limited by the economies of scale that currently limit small capacity liquefaction. If such advances occurred, liquefaction might be done at the service stations, or even within homes.

Some remote and/or rural distribution will retain small capacity electric grids because the capital cost of small capacity hydrogen pipelines may push infrastructure cost trade-offs toward electricity distribution. But, increasingly, large capacity gaseous hydrogen storage will provide continental energy supply security. The inherent ease of storing gaseous hydrogen will provide access to the full range of energy sources. Isolated locations will be able to generate indigenous hydrogen for applications such as farm tractors. Large scale generating stations will be able to operate with steady outputs without requiring load management technologies. Intermittent sources can deliver their energy to the system on demand and then be sent by pipeline to large storage tanks. Electricity, of course, is not storable, although someday it might be possible that some energy storage nodes might employ superconductor technologies using high temperature superconducting materials.

Transformer Technologies

Finally, let's discuss the distant future from the perspective of transformer technologies in order to examine energy source options in two groupings, "reliables" and what we can call "whimsicals."

I think of reliables as sources, like those that feed most of today's electricity generating plants, including coal, uranium, and hydraulic power, that deliver energy whenever called upon, independent of time of day or weather. Sometimes, as in the case of electricity generation, this means at almost constant output. At other times, as in the case of the family automobile, where oil is the fuel source, it means when we want to drive somewhere or want an extra shot of power to accelerate away from the stoplight.

In contrast, whimsicals are sources that deliver their power on their own schedule, which may or may not be when it's needed. They are typified by many renewables that deliver energy when the sun shines, or the wind blows, and so on, but not otherwise.

Although both reliables and whimsicals will contribute during the transition to a fully developed hydrogen age, reliables will always dominate. And among reliables, I expect nuclear fission will be preeminent—at least during the first half-century of the fully developed hydrogen age. Thereafter, it gets harder to be sure. For example, by 2150 it is conceivable that controlled nuclear fusion might have not only become feasible but might come to dominate energy sources.

Considering what will come to be the dominant reliable, one advantage of nuclear fission, especially during the transition to the fully developed hydrogen age, is that its energy product can be rapidly swung from producing electricity to producing hydrogen via electrolysis. To facilitate high efficiency production of both currencies during the transition era, both ac and dc generators can be mounted on a single shaft. This configuration can continue into a fully developed hydrogen age because other reliables, for example, high temperature reactors and fusion, will almost certainly be dedicated to hydrogen production. But we will always need sources with the capability to quickly "swing" between hydrogen and electricity.

The capital cost of nuclear fission will have dropped significantly—especially compared with that of the then-dinosaur-technology coal-fired generation. (As one example, today the capital costs of Advanced Candu Reactors are in the range of $1000 per kilowatt [kW]—about the same as coal-fired plants.) But since the operating cost of a nuclear power plant will always be a small fraction of that for a coal-fired power plant, the energy currencies from nuclear plants will be lower.

Other reliables are likely to play smaller roles in the transition to hydrogen. Natural gas will be used to manufacture hydrogen using "advanced" steam methane reforming (SMR) processes for much of the twenty-first century. The heat for SMR will be provided by nuclear sources, not by natural gas. The requirements for sustainability will require sequestering the waste product—carbon dioxide. This means nearby sequestering sites, leakproof over at least 1000-year time frames, must be available. This need will ultimately place a cap on hydrogen production from natural gas.

When used as energy sources, coal and perhaps oil will become illegal or irrelevant or both. These twentieth century hydrocarbon sources will retain high value roles for things like medicines, materials, lubricants, perhaps even foods. This last might startle. But I expect clever people will learn how to make high quality "simulated" salmon, beef, and tomatoes from oil and coal.

Turning to the whimsicals, the rate at which these renewable resources are harvested must be much less than the replenishing rate by at least an order of magnitude. This consumption–replenishment ratio is too often ignored by today's renewable proponents—who fail to recognize that the patterns of environmental damage caused by overharvesting renewables such as fish, fresh water, and forests will carry over to the environmental damage wrought by overharvesting renewable energy.

The hallmark of nature's mechanisms for delivering renewable energy is low energy density. Nature helps mitigate this limitation in one case, hydraulic, by gathering the energy of diffuse rainfall with her streams, lakes, and rivers. But for almost all other renewables, in most cases, disperse energy delivery dominates. A lightning bolt is one case where nature deviates from her general pattern of diffuse energy delivery. This exception that proves the rule simultaneously demonstrates a second observation: "Nature is kind." What if all renewable energy came in lightning bolts?

The magnitude and location of harvested whimsical energy is capped by a large surface area requirement—whether land or ocean. Moreover, as the twenty-first century unfolds, the use of global surface areas for human development will be even more constrained than today. The magnitude and location of harvested energy will, increasingly, be constrained by the intrusion upon other renewables for which, unlike energy sources, there are no alternatives, such as forests, freshwater, wildlife, and food. Still, with respect to whimsicals, the fully developed hydrogen age will bequeath a major benefit by releasing one of today's critical constraints. Coping with intermittency will no longer be an issue, a consequence of pipeline networks and national storage reservoirs for gaseous hydrogen.

The global need for a transition to hydrogen is critical. The fundamental ideas defining this transition are simple. The lack of widespread understanding of both these facts is stupefying. The dithering is scary. The promise, brilliant.

CHAPTER 3

Prospecting the Future for Hydrogen Fuel Cell Vehicle Markets

Kenneth S. Kurani, Thomas S. Turrentine, Reid R. Heffner, and Christopher Congleton

As there are currently no retail markets for either hydrogen transportation fuel or fuel cell vehicles (FCVs), any discussion of such markets necessarily prospects the future. Such a task is inherently uncertain—many forecasts have been wrong even in mature markets. We undertake this risky enterprise by framing the discussion of future markets for hydrogen and FCVs around two questions. First, what is the history and future of mobility? Second, within this future, why would anyone buy an FCV?

We address the uncertainty of predicting the future by grounding our efforts in a theory of the development of late-modern societies and the related long-term deployment of the infrastructures such societies build to support themselves. We will present historical data on the sociotechnological infrastructures that the modernizing United States built during the twentieth century. Such histories could also be constructed for much of the former British Commonwealth, Europe, Japan, and more recently much of the rest of Asia. We would argue that modern societies are unlikely to deviate from these historical trajectories except in the case of catastrophic events.

The three infrastructures we address are automobility, electricity, and information. These are not the only supporting infrastructures of modernizing societies, but they are three of the most relevant for this discussion. These infrastructures are built to overcome constraints on two fundamental processes—mobilization and globalization. We are motivated primarily by the potential changes that fuel cells and electric-drive technologies bring to automobiles, trends in consumer use of electricity during travel and in their vehicles, and overall trends in consumer lifestyles.

We suggest that the next supporting infrastructure built by modern societies will be a system that fully integrates automobility, electricity, and information. This will be accomplished in part by transforming the current design and role of automobiles as primarily mobility tools. In a technological sense, automobiles will become integrated mobility-electricity-information platforms; in a behavioral sense, they will become "mobile activity locales." One of the behavioral and technological integrations is mobile electricity—the integration of electric-drive, energy storage and delivery, and mobility technologies such that it is possible for the vehicle to deliver electricity for nonpropulsion uses wherever it is, whether stationary or in motion.

Based on this, we argue that in the future FCVs can gain competitive advantage if hydrogen and fuel cells are the best energy carriers and converters to power integrated mobility-electricity-information platforms. (Throughout this discussion we assume FCVs will be directly fueled by hydrogen.) FCVs may also be afforded further competitive advantage by policies that are sensitive to the new role of automobiles as mobile activity locales and create socially sanctioned rewards for progress toward the collective benefits that are the real goals of a transition to hydrogen. While other energy technologies can also provide mobile electricity, FCVs combine mobile electricity with other advantages. Compared to batteries, fuel cell systems can provide more power and more energy; compared to combustion engines, fuel cell systems produce zero local emissions and less noise.

Though societies build a variety of infrastructures, the social sciences have paid less attention to those infrastructures than one might think. The social sciences have paid surprisingly little attention to the automobile itself. Sheller and Urry (2000) note, "... the social sciences have generally ignored the motor car and its awesome consequences for social life." Rosa *et al.* (1988) provide a sociological perspective on energy use: "Energy, though fundamentally a physical variable, penetrates significantly into almost all facets of the social world. Lifestyles, broad patterns of communication and interaction, collective activities, and key features of social structure and change are conditioned by the availability of energy, the technical means for converting energy into usable forms, and the ways energy is ultimately used." We argue that the relationship between society and its infrastructures is less deterministic and more dialectical than this. We argue that the availability of mobility, energy, and information is also conditioned by social structure and change. Moreover, review articles conclude that the "sociology of energy use" has focused on energy use in buildings and has largely ignored transportation energy (Rosa *et al.*, 1988; Stern, 1992; Lutzenhiser, 1995).

Why is it important to place this discussion of hydrogen and FCVs into an explicit social science context? Rosa *et al.* (1988) notes that initial research in the area of household energy use "was guided by the singular assumption, derived from an engineering perspective, that household energy consumption could easily be explained by physical variables." This assumption was exposed as false early in the 1970s by simple research on energy use that included lifestyle and cultural variables.

We apply concepts from sociologist Anthony Giddens' (1984, 1991) "structuration approach" to provide the overall framework in which we evoke an image of the future of mobility and thus the context into which hydrogen and FCVs may enter. We use this framework to deepen social science theorizing about the future of mobility and thus to provide social and lifestyle variables for the discussion of transportation, energy, and the environment.

The encounter between transportation innovations and lifestyle goals of individuals and households provides a rich field of research. Our approach is to ground research in the lifestyles and goals of respondents while employing multiple, staged research methods. This approach (as applied to the case of battery electric vehicles) is described elsewhere (Turrentine and Kurani, 1998; Kurani and Turrentine, 2002b).

Many transportation and energy researchers may view it as odd—even irritating—to encounter sociological theory in a discussion of something like hydrogen FCVs. Most energy and transportation studies focus on technical and market trends during the past few decades, and all within the context of a fully industrialized, automobilized, market-based society. In contrast, historical sociology has drawn much of its theoretical inspiration from analyzing the forces underlying the transformation of traditional societies into modern societies over a period of several hundred years. How is this "grand view" of modern society useful to transportation energy research and understanding the future of hydrogen FCVs? A partial answer is that the grand view allows us to see automobiles, electricity, and information technologies as recent variables in a longer process of lifeway transformation and not as essential or originating conditions.

Mobility is tied up in a broad set of dynamics that characterize modern societies and that contrast with traditional societies whose institutions promote replication and homeostasis. Of particular theoretical interest, Anthony Giddens (1991) attributes much of the dynamism of modernity to "... the separation of time and space and their recombination in forms which permit the precise time-space 'zoning' of social life." What facilitates this zoning? Two basic processes are mobilization and (following from it) globalization. "Globalization" includes the destruction of local measures of time and space, and their replacement with universal measures. "Mobilization" is understood to be the spread of mobility throughout a population; in some sense, it is the democratization of mobility. In this view, the approach toward universal automobile ownership in modern societies is not merely a profound technical and economic transformation; automobility is itself one tool for achieving the time and space zoning of social life in modern society. Under modernity, identity has become "auto-mobile."

A History and Future of Mobility

We explore how FCVs might fit into a future much larger than the automobile market—and go beyond concepts such as automobility. We define "automobility" as personal mobility that allows individuals to move in a self-directed fashion through the space and time of their daily lives. Over the past century, we have built interconnected sociotechnological systems to support and provide automobility. These systems include automobiles, roads, and fuels.

The future we describe is not predestined by this history. While the processes behind these trends are central to modernizing societies, they are also the product of human decision making and social forces. The future we evoke in this chapter will help researchers, policymakers, and ordinary citizens calibrate ideas about future energy use and lifestyle as well as illuminate potential market pathways for hydrogen and FCVs.

How to Represent the Future?

Often the main assumption in studies of technology adoption is technological determinism—that technologies such as automobiles, electricity, and telephones have steered history. We do not follow this view. We argue that successful technologies are those that are in synchrony with the major trends of history. This is our hypothesis about FCVs and mobile electricity: that these technologies might succeed because they follow the main trends of history.

In this chapter we argue that for hydrogen and fuel cells to compete in a future automobility market, they must be integrated with mobility and information systems. Mobile electricity and communications extend the choice framework, especially in terms of time-space zones and personal and social investments in those zones, and emphasize choice mechanisms over localized practices. All locations become globalized and vehicles become addresses.

While automobility systems are central to our thesis, these systems are augmented by communication systems that give "mobility" to information and thus can allow a person to create action at a distance. Further, the capabilities of mobility and communication systems are multiplied by energy systems so that they may move more people, more goods, and more information faster and more broadly. All three of these sociotechnological systems are means to overcome constraints on people's freedom to choose. Hägerstrand (1970) characterized these constraints in terms of capability, authority, and coupling. He sees these constraints as shaping the possible activity space through which a person can and may move. In particular, capability constraints "... arise from biological requirements and the tools available to an individual to mediate time and space."

Before "realists" dismiss this exercise as futuristic optimism, we note that not everything we explore indicates a rosy picture for hydrogen, FCVs, and the future of automobility. We are not talking about progress per se, but rather we try to situate FCVs within long term trends and within feedbacks between culture and technology. We note that competing automotive technologies will also offer mobile electricity; FCVs will not have this market to themselves. Moreover, the lifestyle trends we examine appear to run counter to the hopes of many hydrogen and FCV proponents for less energy use. These trends also challenge the hopes of "new urbanists" who call for a future of reduced personal vehicle use.

Next, we present data regarding the development of automobility, electricity, and information infrastructures. Their simultaneity in time is not coincidental but is evidence of a modernizing society invoking capabilities that in turn shape modernity. While we will present measures of "things," e.g., vehicles, gas stations, cellular transmitters, and such, over time and space, the point is that these are evidence of new capabilities being spread broadly through the population.

Mobility Infrastructure

Here we examine the development of mobility systems, primarily automobility. The three main supporting physical infrastructures for automobility are roads, vehicles, and fuels. We focus on the roads and vehicles first; motor fuel will be treated in the subsequent section on energy.

Roads

Until well into the twentieth century, the majority of roads in the United States were little more than dirt tracks. The 260,000 miles of railroad in the year 1900 was more than double the 125,000 miles of road that had any type of improved surface (U.S. Department of Transportation, 1976). Only two percent of those were paved with asphalt or concrete—and the majority of these were in urban areas (American Association of State Highway Officials, 1953). In short, the only national transportation network capable of comparatively high speed and volume transportation services at the turn of the nineteenth to twentieth centuries was the railroads. The railroads were an important step forward in mobility; however, the capability of the railroads to facilitate automobility was limited.

A great step in road building in the United States was the construction of the national network of limited access highways. That system, known as the National System of Interstate and Defense Highways, was initiated with legislation signed by President Eisenhower in June of 1956. It is now essentially complete, consisting of over 42,000 centerline miles. No new interstate centerline miles are planned; any additions of lane-miles, i.e., road widening, will be approved locally.

While the construction of the national interstate highway system was one great step in twentieth century road building in the United States, the improvement and paving of millions of miles of other roads is another step that further facilitated the mass marketing of automobiles and thus the spread of automobility. Between 1941 and 1995, total centerline miles of public roads (paved and unpaved) increased from 3.3 million to 3.9 million—a modest 18 percent increase over 54 years. On the other hand, centerline miles of paved public roads increased from roughly 600,000 to 2.4 million—a 120 percent increase.

Vehicles

From the year 1900 to 2000, the population of automobiles in the United States grew from essentially zero to a level where there were more automobiles than licensed drivers. According to Stilgoe (2001), about 300 automobiles were operating in the United States in 1895; in 2001 there were nearly 250 million light-duty vehicles (U.S. Department of Transportation, 2001). Data on all highway-licensed motor vehicles except for motorcycles and motor scooters for the period from 1900 to 2001 are plotted in Fig. 3-1, as are data on resident population and licensed drivers.

Early in the twentieth century, automobile ownership was concentrated in urban areas—partly as a function of wealth and partly as a function of the availability of paved roads. According to McShane (1994), the

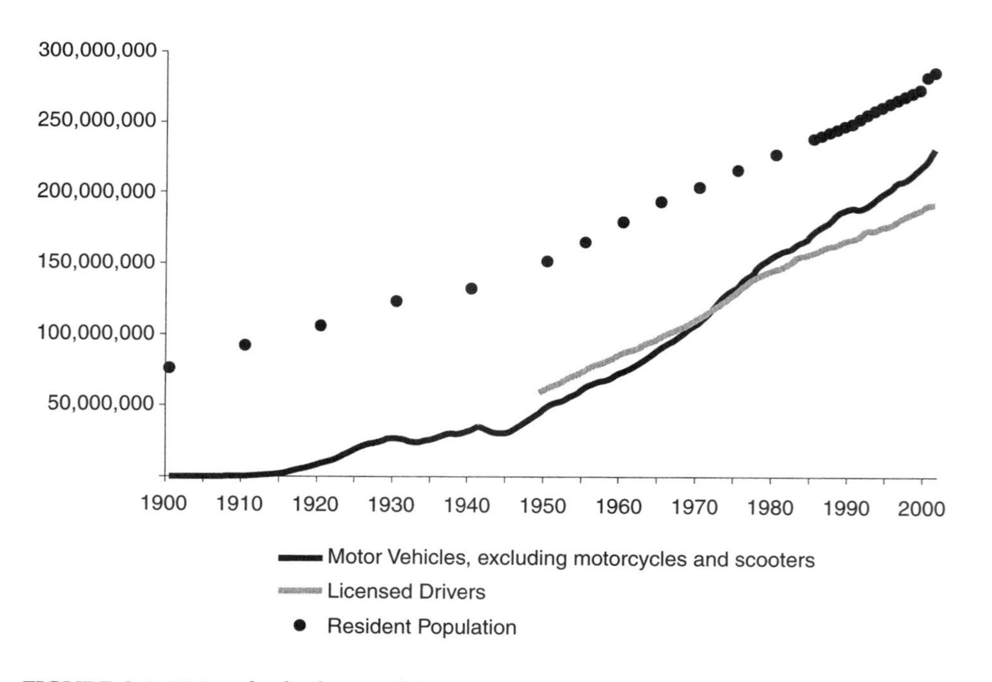

FIGURE 3-1. U.S. vehicle, licensed driver, and resident populations during the twentieth century. *Sources:* Vehicles and licensed drivers: U.S. Department of Transportation (1996a and 1996b to 2001); residents: U.S. Department of Commerce (2002b).

highest per capita concentrations of automobiles were to be found in the cities of Washington, DC, and New York. Automobiles and automobility have been spreading ever since.

By 1960, only 21.5 percent of U.S. households owned no motor vehicles; by the year 2000 this had dropped to 9.5 percent (Davis and Diegel, 2002). In fact, by the year 2000, most U.S. households (56.8 percent) owned two or more vehicles. By 2001, 22.4 percent of U.S. households had more vehicles than drivers (Hu, 2003).

Still, automobility is not universal in the United States. In 2001, 7.9 percent of U.S. households did not own a motor vehicle (U.S. Department of Transportation, 2003). We infer that in many cases this is not voluntary but rather caused by poverty. Households with incomes less than $25,000 (in 2001 dollars) account for 20.3 percent of households that own no vehicles (U.S. Department of Transportation, 2003). As a modern society that sustains and is sustained by automobility, we argue that our society obligates itself to ensure that all its citizens have the means to participate in social, civic, and political life.

Energy Infrastructure

We address two energy infrastructures—the retail gasoline network and the national network for distributing electricity. The latter has powered some new automobiles introduced in response to air quality requirements and serves as one alternative model of a hydrogen distribution network. At the start of the twentieth century, people consumed limited quantities of locally produced energy. Throughout the century, systems were built to provide more people with increasing amounts of energy from an increasing variety of fuels.

Gasoline

Concurrent with the widespread adoption of automobiles and the construction of a nationwide system of paved roads, a retail fueling network was built for those vehicles. The growth of a gasoline retail network specifically for automobiles began in the first or second decade of the twentieth century—there are several competing claims for when the first "service station" was built. Prior to the development of gasoline stations, gasoline was sold in cans that could be bought at grocery stores, lumber yards, and coal merchants, as well as carriage, blacksmith, and livery shops. By 1920, gasoline pumps were evident throughout North America.

Counts of retail gasoline outlets are neither consistent nor precise since the reality of gasoline retailing has shifted from sales of canned gasoline at retail outlets not specialized to service automobiles, to stations dedicated to sales of fuels and services intended primarily for automobiles, and now back toward more generalized stores selling a variety of food, beverages, and other goods and services. Lately, this trend has included the sale and branding of gasoline by large general retailers and grocery stores. We provide a composite estimate of the changes in gasoline retailing in the United States during the period from 1920 to 2003 in Fig. 3-2. There were about 12,000 retail gasoline stations in the United States in 1920, increasing to 124,000 by 1930 and peaking at over 225,000 around 1970, followed by a steady decline (Jakle and Sculle, 1994 and U.S. Department of Energy, 2003). The decline in retail locations results in an increase in the average amount of gasoline sold per retail location as gasoline use continues to rise. Despite the reduction in the number of retail gasoline outlets in the United States during the 1990s, gasoline stations remain ubiquitous.

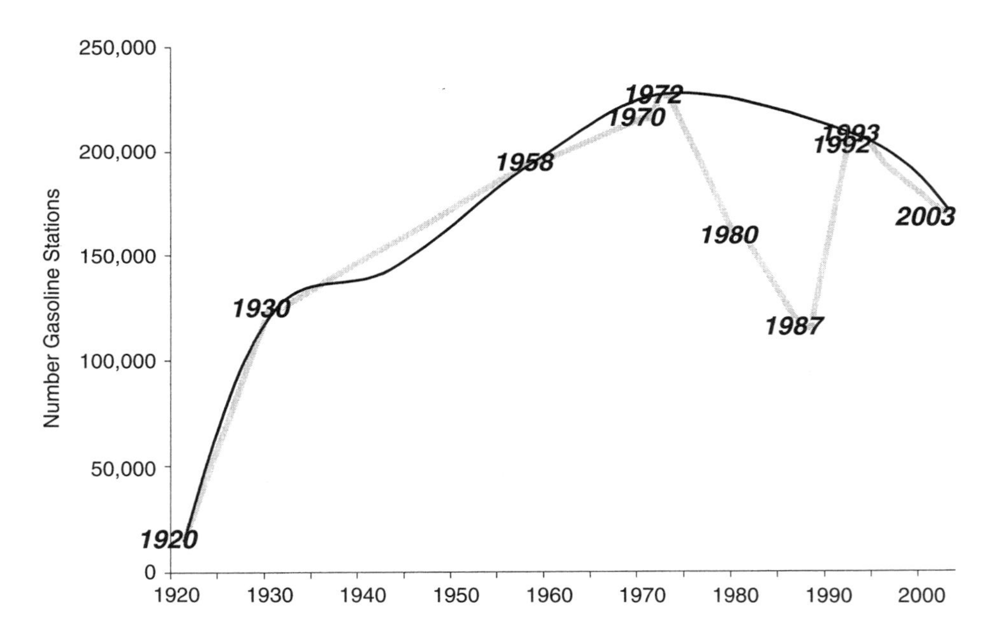

FIGURE 3-2. Composite estimate of long-term trends in gasoline retail infrastructure in the United States, 1920 to 2003. *Sources:* 1920 to 1987: Jakle and Sculle (1994). 1992 to 2003: *National Petroleum News,* cited by U.S. Department of Energy (2003).

Note: Data points in Fig. 3-2 are shown by a label indicating their year. The result of treating all points as if they measured the same thing is the implausible history shown by the gray line. We argue that the black line is a more plausible history, based on the assumptions that U.S. Census data prior to the early 1970s and NPN data in the 1990s are reasonably accurate counts of the locations at which retail customers could buy gasoline and that growth of the retail gasoline network stalled during the 1930s and early 1940s as did sales of automobiles.

Electricity

The beginnings of the social and technological transformations of electrification predated those of automobility by only a few years. Electrification facilitated increases in the variety of fuels consumed and in the amount of energy consumed—directly and indirectly—and allowed for the distribution of energy locally and across long distances.

Thomas Edison built his first electric power plant in 1882. It produced direct current electricity. Because of transmission losses, the users of its electricity were limited to businesses and residents near the power station. The electricity was used primarily for incandescent lighting. Today nearly 100 percent of American residential and business locations are served by alternating-current electricity. By the year 1930, approximately 90 percent of urban residents in the United States were served by electricity, while only 10 percent of rural residents were so served (New Deal Network, 2003). Through the efforts of the federal Rural Electrification Administration, the percentage of rural residents who had electricity increased to 25 percent by 1939. Farm electrification lagged, with only 11.6 percent electrified in 1935. By 1953, nine out of ten farms were served by electricity, and by 1963 farm electrification was nearly universal (97.9 percent). See Fig. 3-3 for a summary of farm electrification service levels. Data from the 2001 American Housing Survey indicate that greater than 99 percent of occupied American residences had electric service of some kind (U.S. Department of Commerce, 2002a).

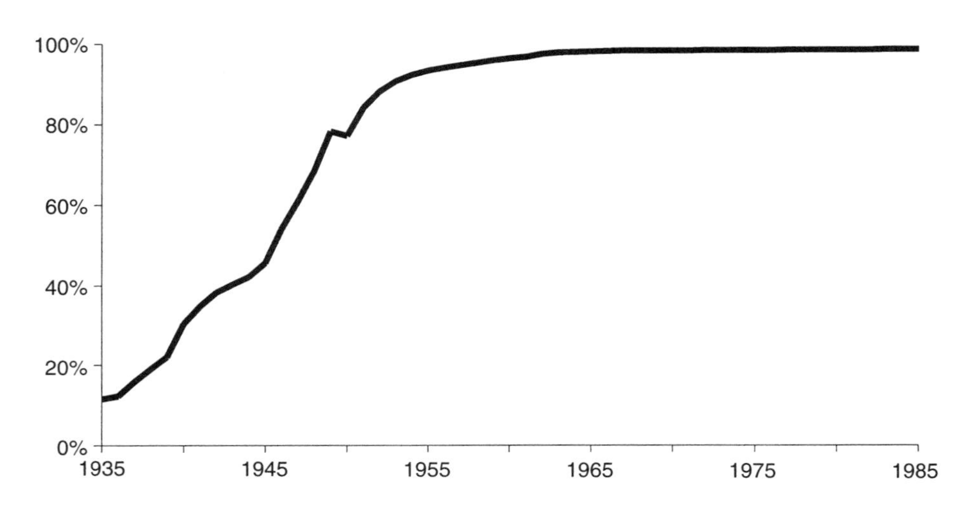

FIGURE 3-3. United States farms with electricity service, 1935 to 1985. *Source:* U.S. Department of Agriculture (1985).

Information Infrastructure

Information systems—by which we mean an interrelated suite of communication and computation technologies—are the third infrastructure we consider. In addition to the spread of information services across the population, the other important development during the twentieth century was the advent of wireless information systems. First, the country was hardwired for telephony, and then systems of wireless telephony were built. First, the internet was built on a hardwired infrastructure. Now increasingly it is based on wireless networks. These trends point to a future in which information continues to be untethered from location.

Telephony

Alexander Graham Bell's first phone call to his assistant in the next room was made on March 10, 1876. By 1920, 35 percent of U.S. households had telephones. Household access is now nearly universal. The Federal Communications Commission (2003) reports 97.6 percent of U.S. households had a landline telephone in the year 2000 (see Fig. 3-4).

Significant differences in residential landline telephone service existed between urban and rural populations for much of the twentieth century

FIGURE 3-4. Percent of U.S. households and farms with landline telephones, 1920 to 2000. *Source:* Households: Federal Communications Commission (2003). Farms: U.S. Department of Agriculture, Rural Electrification Administration (1985).

and continue to exist across the population by income. In 1940, only about two-thirds as many farms had landline telephones as did all U.S. households. This gap between urban and rural residents appears to have closed by 1980. While residential access to a telephone is now nearly universal, discrepancies by income persist. As late as March 2000, only 87.5 percent of households with incomes less than $10,000 (in 1984 dollars) had residential telephone service (Federal Communications Commission, 2001).

The wireless phone network and the cellular phones that use it got their commercial start in the United States in the early 1980s (see Fig. 3-5). In 1984, there were 346 cellular transmission sites in the United States and 91,600 subscribers. It took only 20 years for nearly half of Americans to access the cellular phone network. By December 2002, there were nearly 140,000 cellular transmission sites and over 140 million U.S. subscribers. The change in geographic distribution, and in particular the increase in density, of the cellular transmitter sites is shown in Fig. 3-6.

The cellular phone system has not replaced the landline phone system, but supplemented it. Still, for many people cellular telephony has transformed communications: A telephone is no longer a location to be sought out, but a personal accessory.

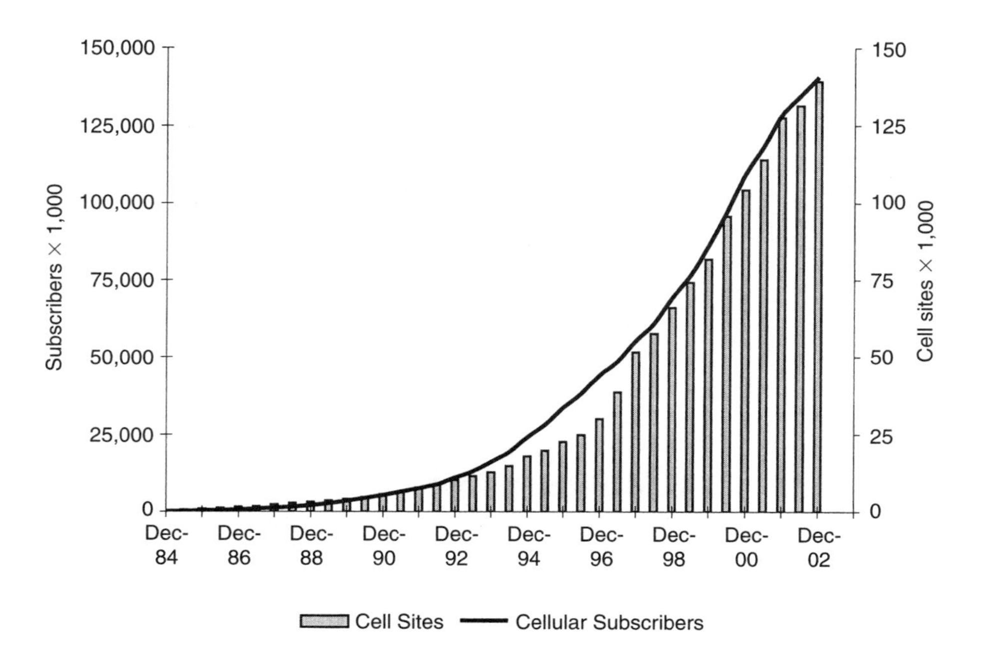

FIGURE 3-5. Number of cellular transmission sites and cellular phone subscribers in the United States. *Source:* Cellular Telephone and Internet Association (2003).

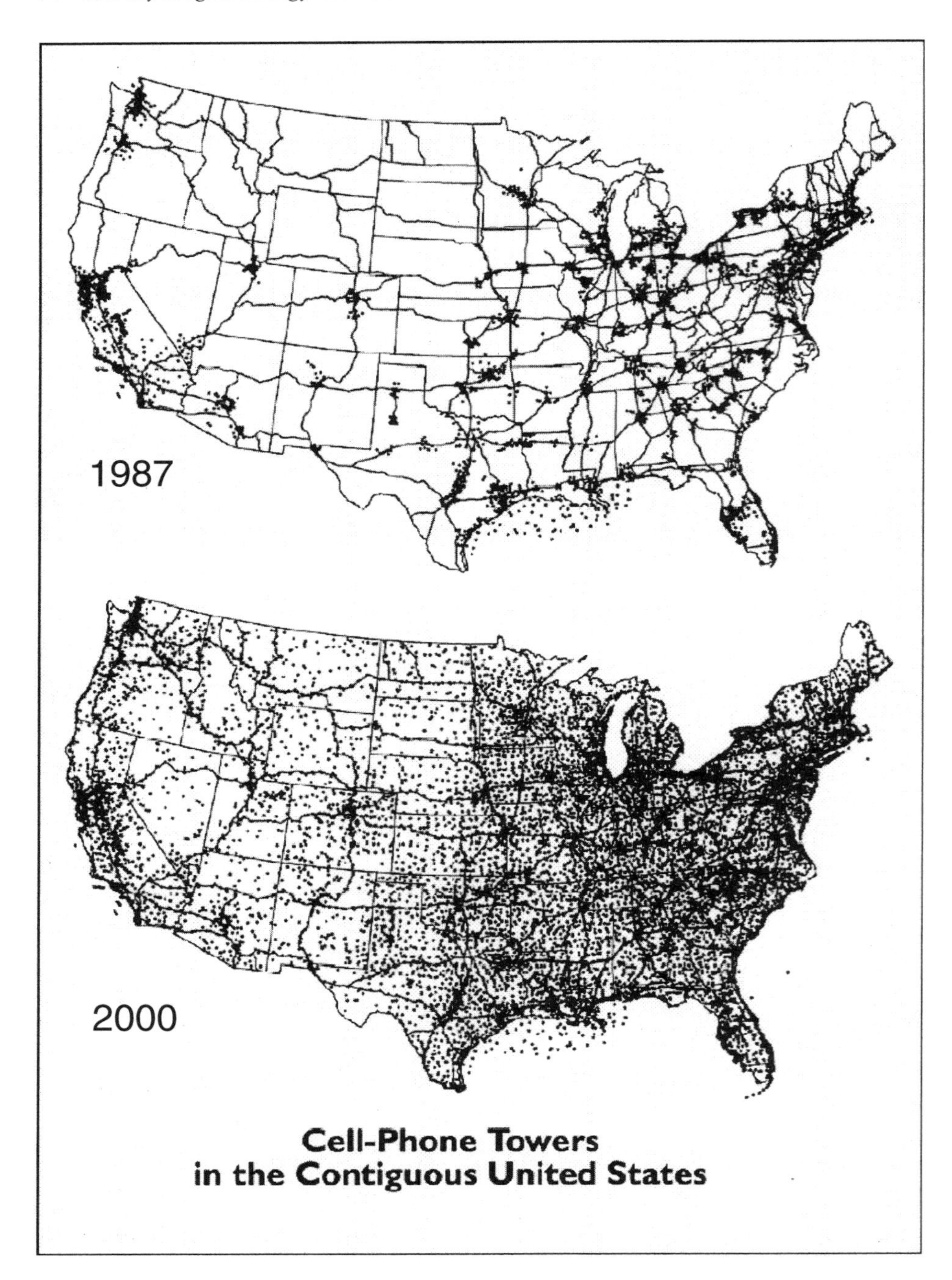

FIGURE 3-6. Change in geographic distribution of cellular transmitter sites in the United States, 1987 to 2000. *Source:* Winkle (Copyright American Geographical Society, 2002).

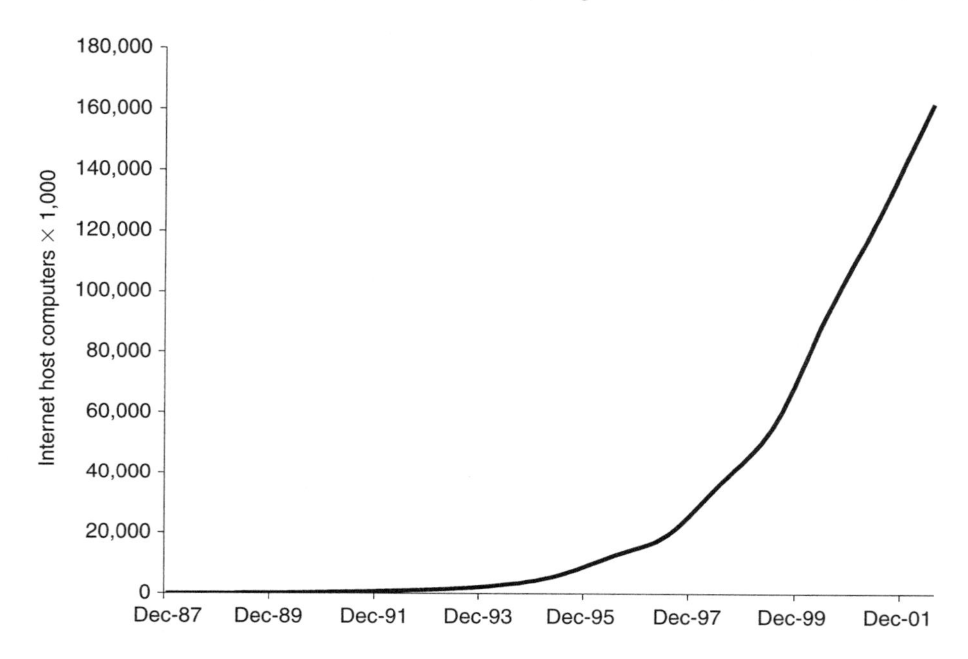

FIGURE 3-7. Internet host computers, international. *Source:* Internet Software Consortium (2002).

The Internet

The Internet operates largely on a backbone of dedicated, high speed, hardwired communications "pipelines" and specialized computers—routers—to direct the flow of information. In 1981, the precursor to the Internet consisted of 213 interconnected computers. This hardware system grew slowly through the 1980s, tripling in size over 10 years. Only in the early 1990s did the system begin expanding rapidly; there were an estimated 162 million host computers by the end of 2002. This trajectory is plotted in Fig. 3-7.

Estimates of the growth in the number and proportion of the adult U.S. population who are Internet users are presented in Fig. 3-8. In only 7 years, the percentage of the adult U.S. population who use the Internet grew from 9 to 66 percent.

Another way to measure the growth of the Internet is by the increase in the number of Internet protocol (IP) addresses. Devices connected to the Internet are assigned an IP address. The current common form is an IP version 4 address, with 4 billion possible addresses. Because of growth in the number of host computers, routers, and users as well as inefficiencies in how IP addresses were first allocated, some industry analysts estimate all available IPv4 addresses will be allocated sometime in 2005. The number of next generation IPv6 addresses is estimated to be greater than 35 trillion,

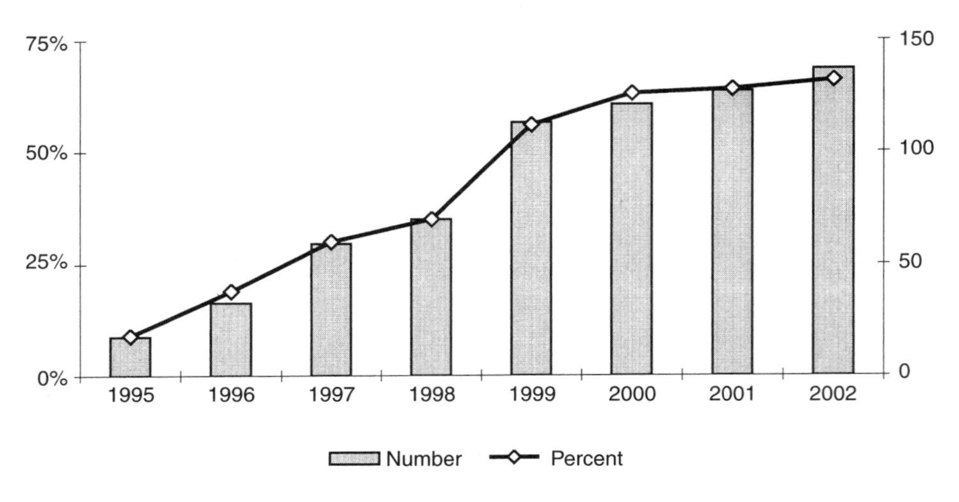

FIGURE 3-8. United States adult Internet users. *Source:* Harris Interactive (2002).

allowing virtually every electronic device to have its own IP address. IPv6 addressing is one standard that makes it practical for automobiles and their occupants to communicate with a wide variety of vendors, information sources, other vehicles, and people.

Wireless Networking

Another recent development essential to increased communications is wireless networking. Wireless networks are appearing in homes and businesses, and as well as restaurants, airports, movie theaters, and other locations that offer public access to wireless networks—some for a charge, others for free. Data on the initial proliferation of public Wi-Fi (wireless) networks and users are shown in Fig. 3-9.

Integrating Infrastructures

Our thesis is that a next step in the building of infrastructures for modernizing societies will be the integration of automobility, electricity, and information infrastructures. This process has started in a number of ways—already there are systems that integrate pairs of these infrastructures. We have discussed combining mobility and communications in the example of cellular telephony. Mobility and electricity are already being more closely integrated, so the complete integration of all three appears technically possible.

Automobility and Communications

Wireless network technology is appearing in vehicles. Wireless networks are being used to integrate communications with a driver's cellular telephone.

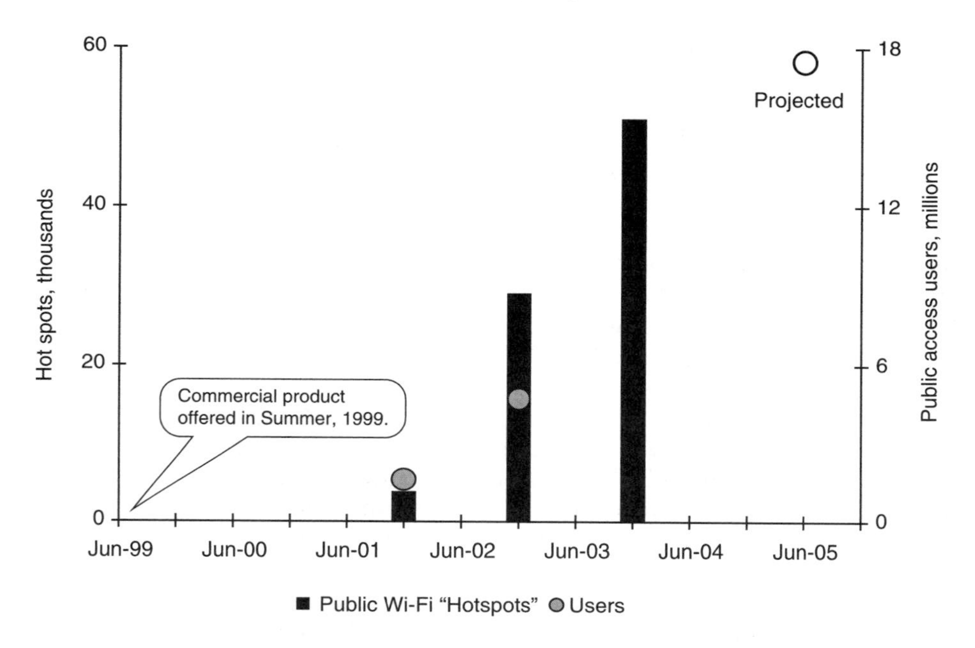

FIGURE 3-9. U.S. public Wi-Fi networks and users. *Source:* Tedeschi, B. 2003. Eating Out and Logging On. *New York Times*, June 14.

Wireless networks may also handle onboard communications, for example, replacing hardwiring between sensors and computers. This could reduce production and maintenance costs. Wireless networks could handle communications between vehicles and a variety of stationary or mobile devices, depending on distances and data quantities. Such communications can and will facilitate payment of road and bridge tolls, fuel transactions, and other purchases. Vehicle safety systems could be built around vehicle-to-vehicle communication networks, for example, to improve adaptive braking systems. Improved wireless networking can also make full-screen, high-speed Internet access available in automobiles, improving on the few lines of text available on existing cellular phones.

Mobile Electricity

Mobile electricity is already happening in a number of ways. Some of these represent important markets for auxiliary electric power devices such as portable generators; some are extensions and expansions of long available, but heretofore minor, capabilities, such as the 12-volt "cigarette lighter" outlet in vehicles.

The number of 12-volt electric outlets in automobiles has increased in recent years, particularly in light trucks. Most small and midsized SUVs have outlets in the front and rear. Larger SUVs have three zones of 12-volt outlets—front, rear, and cargo area. General Motors' Hummer H2 comes with a total of five such outlets. Newly redesigned smaller SUVs and mini-vans are also including multiple 12-volt outlets in three zones of the vehicle. This proliferation of electric outlets is relatively new in smaller vehicles and may indicate that consumers are demanding greater onboard electric power.

Portable generators powered by gasoline fueled internal combustion engines (ICEs) are another way that people have been making electricity mobile. The construction trades, recreational vehicle owners, festival promoters, and many others have used these ICE-powered devices to provide electricity off the electric grid. But these mobile electricity applications are marginalized by noise, localized exhaust emissions, and weak or absent integration with vehicles.

The converse of taking electricity off the grid is supplying electricity to the grid. Building owners have used standby generators, fuel cells, and battery banks to provide backup electric power to applications that are especially sensitive to disruption. These include hospitals, attended care facilities and health clinics, banks and credit data processing centers, and even homeowners.

Combining mobility and electricity raises the question of which of these types of applications might an FCV fulfill. Further, the prospect of mobile electricity raises the question of what ways might lifestyle activities and business operations change if high power mobile electricity doesn't require a separate portable generator but is fully integrated into every vehicle sold.

Evoking the Future

As the data of the previous sections have shown, we have built and continue to build systems of automobility, energy, and communications. Over the twentieth century, we extended and intensified physical access to the landscape. Over much the same time, we built a hardwired network of telephony for real-time communication. In the latter part of that century, we began to spread communications capability more finely across space and time by freeing telephony from location. We are now pushing further to spread access to information to "everywhere" through wireless networking. And while once abundant electricity was also tied to location, the capability to take relatively large amounts of clean, quiet electricity and spread it across the landscape in the form of mobile electricity fits this historical pattern—even if this wasn't the reason we started to contemplate FCVs in the first place. The integration of all these—automobility, electricity, and information—will result in the transformation of the automobile into what we are calling the mobile activity locale.

Giddens (1984) conceptualizes "locales" not primarily in terms of a place but in terms of the use of a place as a setting for social interaction. The automobile as integrated mobility-electricity-information platform becomes a mobile activity locale whether it is stationary or in motion. A mobile activity locale is mobile both in the sense that some activities may be undertaken while it is in motion and in the sense that it facilitates the convening of activities in novel geographic locations. If your car is fully equipped to be your workplace, then your workplace isn't permanently defined by a street address. Your business addresses may be a URL in the virtual world and a GPS transponder in the physical world. For some businesses, the only practical reason for a street address or post office box may be legal requirements that it be incorporated some "where." For parents busy chauffeuring children, the automobile becomes a mobile study hall, video arcade, changing room, dining room, and bedroom. The automobile becomes a mobile locale for activities formerly reserved for offices, homes, schools, restaurants, and other stationary locales.

This then is the image of the future we are evoking, the future context we believe FCVs will enter. The processes of mobilization and globalization, and the sociotechnological systems that modern societies build to sustain those processes lead, to a world in which each of us, increasingly, must ask ourselves the following question: What will we do—when we can do anything, anywhere, anytime?

Why Would Anyone Buy a FCV?

In short, the reasons to buy an automobile will shift from primarily transport between stationary locales to new lifestyle and work patterns incorporating mobile locales—patterns that we see are already developing. Many workers currently find work-related activities intruding into what were formerly nonwork times and places, using cellular phones, laptop computers, and other mobile information tools to connect with their offices. Spatially separated household members use cellular telephones to attempt complex "relevant time" coordination of their schedules.

One of the compelling pieces of evidence that the automobile of the future will be a mobile activity locale is that automakers are already redesigning their vehicles to be such. While mobile office features are not yet standard offerings in current vehicle models in the United States, mobile electricity and mobile information technology are showcased in select concept cars. For example, DaimlerChrysler's Dodge MAXXcab concept vehicle shown at the 2000 North American Automobile Show included a built-in laptop computer, Internet access, a DVD-entertainment center, and voice recognition controls, all powered by a next-generation 42-volt electric system. Suzuki showed a minivan at the 2003 Tokyo Automobile Show designed to be reconfigured as an office, in part by transforming its instrument panel into a table. More mundane, and therefore more important, automakers now routinely offer a number of standard or optional features that are aimed at the multiple activities already being carried out in vehicles and that hint at future capabilities. Examples include rear-seat DVD entertainment systems, multizone heating and cooling, and pet-friendly restraints. We've already discussed the proliferation of 12-volt outlets in cars and trucks. In addition, Toyota and GM now offer a 115-volt, U.S.-household style, two-prong electric outlet in a few models of light-duty vehicles.

A New Product

If we sell FCVs as mere replacements for today's ICE-powered vehicles (ICEVs), their market will develop slowly, if at all. In particular, the driving range of hydrogen FCVs will likely be limited, compared with ICEVs, for some time. Thus, FCVs would seem to suffer some of the same perceived barriers as do battery electric vehicles—limited driving range per refueling and high cost. FCVs may also suffer a constraint that battery EVs do not—the lack of home refueling. The more FCVs are portrayed as similar to ICEVs, the more FCVs must compete in arenas where ICEVs dominate. That is a daunting challenge given that our society has spent the past 100 years building a system of automobility based on ICEVs. One clear lesson from the last two decades of experimentation and small-scale market launches of a variety of alternative fuel and electric vehicles is that gasoline and diesel are hard to beat—on their terms.

The nature of our approach is to change the terms by proposing a new product category (mobile activity locales) and a new marketing approach (social marketing), so that gasoline and diesel vehicles are not necessarily superior. If we focus on the innovative aspects of FCVs, in particular what we call their "lifestyle attributes," we allow FCVs to compete in arenas where they have advantages. Lifestyle attributes of FCVs are those that open new lifestyle activities for households, activities in which households can invest themselves. These activities create new values for vehicles—values which ICEVs may have a difficult time providing.

For example, the attribute of onboard, clean, quiet, high power electricity allows consumers to use household-like appliances in their vehicles, charge batteries for their electric power tools, take a microwave oven on their picnic, bring their television to the beach, run a business out of their vehicle, or furnish a campsite with lighting and electric heat. As with many emerging technologies, the complete scope of uses cannot be fully predicted. Consumers will be the big innovators here, creating whole new uses (activities) for this new attribute. This in turn will affect the design of all vehicles, and possibly pull different designs of competing technologies into the market. For example, the advent of mobile electricity may pull hybrid electric-ICE vehicles toward larger batteries and more powerful motor/generators. In fact, large-battery, grid-rechargeable hybrid electric vehicles may have two inherent advantages in the short run as mobile electricity sources over FCVs—the high-energy content of gasoline and the existing, ubiquitous network of retail gasoline stations.

Additionally, the green market value of FCVs (and competing alternatives) is a lifestyle attribute in that it symbolizes a new way of life to buyers. Many consumers are seeking a way to maintain their automobility while reducing their impact on the planet and on the health and well-being of others. We have found in interviews with hybrid vehicle buyers that much of the value some of them receive is derived from their sense of being an innovator and from the conversations about their vehicle with others—the sense of being part of a social movement.

How does society ensure that FCVs provide it with the benefits it wants? How do consumers capture FCVs' lifestyle attributes? Conversely, how do they avoid the situation in which the benefits they get and costs they incur are subject to what may ultimately appear to be arbitrary decisions to solve narrow technical problems? One way to avoid such situations is to reposition societal goals, collective benefits, and lifestyle attributes through social marketing. Collective benefits are the subset of public goods that no one gets unless many people—a community, a society—act in concert to acquire them (Kurani and Turrentine, 2002a). Clean air, improved public health, reduced risk of global climate change and war, and diminished and repaired damage to ecological systems are collective benefits. Society requires policies to create a market context that values these collective benefits appropriately. In the past, policies intended to produce collective benefits have included market based incentives, performance requirements, production mandates, partnerships with industry, and differential access to transportation related infrastructure. The latter include electric and natural gas vehicles having access to carpool lanes, preferential parking, and car-free zones. These policies span all levels of governance—federal, state, and local. This creates a rich set of possibilities not only to support technology research and development, but also to create and evaluate a wide variety of socially created benefits and conveniences to the drivers of vehicles that contribute to the attainment of collective benefits.

What about Competing Automotive Alternatives?

Even if FCVs were available in the present, the alternatives against which they would compete are changing. The continued tightening of motor vehicle emissions standards has produced continued progress in limiting emissions of criteria pollutants from internal combustion engines. As FCVs continue in their development phase, the stock of ICEVs will continue to get cleaner. It is less clear that ICEVs will get any more efficient or reduce their greenhouse gas emissions. However, the entry of hybrid electric vehicles (HEVs) into the marketplace may increase the efficiency of the light duty vehicle fleet. Automobile manufacturers continue to plan to increase HEV body style and drivetrain options.

How will cleaner ICEVs and new HEV alternatives change consumer expectations of cars and trucks? It seems plausible to us that over the next several years HEVs will change baseline consumer expectations of automotive performance. Many consumers will come to appreciate the quiet, smooth launch from a stop that can be provided by the electric motor. Many will come to expect a driving range of 500 to 600 miles, rather than the 300 to 400 miles they currently achieve in today's ICEVs. Many will come to appreciate the convenience and lower cost of less frequent refueling. Some will come to expect to be able to choose a vehicle that produces lower greenhouse gas emissions.

FCVs Offer a Unique Bundle of Attributes

One reason we argue that FCVs are a new product is that they represent a unique combination of attributes and performance capabilities. As we have described, competing technologies, including HEVs and ICEVs, could provide some of these attributes and capabilities. However, they cannot provide the combination of the following benefits:

- Clean air, reduced risk of global climate change and war, and eased exploitation of wilderness for energy
- Electric drive feel and new vehicle designs
- Energy from varied fuel stocks and production processes, in particular the potential for carbon free energy
- Automobiles as mobile activity locales facilitating new lifestyles and work structures

FCV and Fuel Infrastructure Performance and Design

The integration of mobility, electricity, and information technology into a single platform will change vehicle and refueling infrastructure design. Some changes will be to basic engineering and design elements of vehicles. If automobiles are going to be transformed into mobile activity locales, they must be capable of providing for the application of electricity to uses other than vehicle propulsion, ancillary lighting, and heating and cooling. Electric outlets and devices must be provided onboard the vehicle.

In addition to simply adding a few outlets, the vehicle itself will change too. If people are going to be doing more things in vehicles, vehicles may get larger (especially smaller ones). People will need room to store mobile appliances and devices, and more space for occupants since they will no longer simply face forward in their seats. People may also desire space that can be configured in multiple ways. This might be accomplished fairly simply with vanlike vehicles. More complex solutions might include the interchangeable bodies that General Motors has suggested could accompany development of their HyWire vehicle—whole different bodies may be swapped on the same platform that houses the basic drivetrain, suspension, braking, energy storage, and conversion systems that power the vehicle.

Relationship between Driving Range and Nontravel Use of Energy

The idea that some energy stored onboard the vehicle might be used for purposes other than vehicle propulsion and the attendant idea that such nonpropulsion uses may promote larger vehicle size have clear implications for driving range (for a given amount of energy stored onboard). Driving range will be reduced by both these things—all else being equal. This creates pressure for all else to not be equal.

It speaks to a possible need for a different hydrogen refueling network morphology than the existing network for gasoline. A refueling network to support the uses of hydrogen powered mobile activity locales may need to be both denser and more extensive. Hydrogen refueling locations may need to be at different types of places. Notably, increased energy use in mobile activity locales and ties between vehicle and home developed by the intrusion of "home activities" into the vehicle may create strong incentives to solve the problem of viable home refueling of FCVs. The "energy station" concept, wherein a large natural gas re-former provides both energy to a stationary base load such as a building and hydrogen for refueling the vehicles of the building's occupants, is another way in which the density, distribution, and location of hydrogen refueling may be different than that of gasoline. In this sense, the hydrogen refueling network might mimic the natural gas network. And certainly the use of onboard energy for nonpropulsion uses provides incentives to increase onboard energy storage and the efficiency of onboard appliances and services.

To provide a practical application of the ideas we discuss throughout this chapter, we look at the implications of mobile electricity for driving range and onboard energy storage goals. The following statement is from the U.S. Department of Energy (2002): "The overarching technical challenge ... is how to store the necessary amount of hydrogen needed to fuel the vehicle for its required driving range (>300 miles), within the constraints of weight, volume, efficiency, and cost." Based on our prospecting of the future, there are several reasons to question the characterization of any driving range target as "necessary" and "required" before integrated analyses of mobile electricity, refueling network morphology, and societal goals are attempted. Keeping in mind that the real purpose of the statement above is to establish a goal for onboard hydrogen storage, we provide reasons why the stated goal is too low and, possibly, too high.

The amount of hydrogen required to drive 300 miles may be too low for two reasons. First, it's too low if we expect people to drive 300 miles and make significant use of onboard energy for nontravel services, i.e., mobile electricity. Second, it's too low if, in the interim while FCVs are still being developed, HEVs shift baseline consumer expectations of driving range up into the range of 500 to 600 miles.

Conversely, a goal to store sufficient hydrogen onboard to drive 300 miles could be too high. What if the refueling network morphology for hydrogen is based on the natural gas or electricity distribution networks, not the gasoline retail network? In particular, if home refueling of hydrogen is possible, then the relationship between onboard storage and refueling is more like an electric or natural gas vehicle than a gasoline or diesel vehicle. And if we learned anything from battery electric vehicles, it is that a 300-mile driving range is not required if (1) the vehicle can be refueled at home and (2) it is one of a variety of travel tools available to a household (Kurani, Turrentine, and Sperling, 1994, 1996). Further, the benefits of mobile electricity may be such that people will accept more frequent refueling—even in a retail fuel network such as the current gasoline network.

Policy Goals and Social Marketing

Hydrogen and fuel cells are means to other ends. Based on the image of the future we develop here, we foresee two problems in achieving those ends. The first is our ability to capture collective benefits. This problem is not specific to FCVs. The second is how (or whether) to market collective benefits of new automotive energy systems. This may also not be specific to FCVs but does point to a need for reanalysis of the likely environmental and geopolitical impacts of both FCVs and their likely competitors. We address this in the following section.

Certainly providing collective benefits cannot be left solely to the market. We have discussed elsewhere (Kurani and Turrentine, 2002a) how collective benefits suffer from the same market failures as do positive externalities—since people can derive benefits for which they do not have to pay, markets will tend to produce too little of any products and services that provide collective benefits. Vehicles providing power to other users and activities could be providing large positive collective benefits. Social marketing might be used to promote the policy goals associated with these benefits, and the sense of participation in creating positive social change (Kurani and Turrentine, 2002a).

The analytical problem created by the transformation of automobiles into mobile activity locales is that it confounds prior evaluations of transportation's environmental and geopolitical impacts. The world used to be neatly—if conceptually—divided into mobile and stationary sources of impacts. Now, cars and trucks are used primarily for mobility; their ancillary energy use for lighting, heating and cooling, and entertainment is generally small in comparison. Mobile activity locales raise the prospects of significant nonmotive energy use in things that may look like cars and trucks but that are used for an expanded variety of activities. Whether mobile activity locales powered by hydrogen fuel cells—or any other energy system—actually create collective benefits needs to be reanalyzed in view of these possibilities.

More energy intensive lifestyles and work patterns could evolve in a number of ways. One example would be the potential for increased duplication of heating and cooling services in vehicles and in buildings. If people cool their cars while using them as an office at the beach, the office in their building may still be cooled. Another example would be the novel introduction of electrically powered services to locales. Campers, who might never imagine they would use a portable generator to electrify their campsites, may find their vehicles' integrated electric generation capability irresistible. Far more prosaic is the possibility of larger numbers of people either preheating or precooling their vehicles. Just what people will do with enhanced heating and cooling capabilities is unknown. What we do know is that such services are becoming universally available in automobiles. The American Automobile Manufacturers Association (1995) reports that in excess of 90 percent of new cars and light trucks now sold in the United States are equipped with air conditioning.

Larger vehicles seem inevitable if people are going to be using their automobiles for more than mobility. People may want to be able to rearrange the space in their vehicles to create different locales; they may simply need more room to carry more appliances, furniture, and other accoutrements. Finally, with the ability to turn any location into a variety of novel locales, it seems plausible that people will travel more, or at least spend more time in their mobile activity locales than they now spend in their automobiles.

A Cautionary, but Motivational, Tale...

We believe the forgoing provides a cautionary but still motivational tale. The fuel cell itself may be a clean and efficient device. However, whether it leads to greater or lesser energy use, greater or lesser emissions of criteria pollutants and greenhouse gases, greater or lesser reliance on petroleum (at least over the short- to mid-term) depends on our ability to foresee and adapt to changes in automobiles and automobility that we believe are all but inevitable. If automobiles are transformed into mobile activity locales, then energy consumption and associated emissions might increase because of more energy intensive lifestyles and work structures, a shift toward larger vehicles, and increases in travel.

Faced with these prospects, integrated analyses are required to judge whether FCVs can actually create progress toward societal goals while further facilitating mobility. We require explicit interaction between market analyses of mobile electricity, FCVs, and their competing alternatives—and analyses of emissions of greenhouse gas (GHG) and criteria pollutants, oil consumption, the geopolitics of oil, development patterns and land use, and wilderness access and road ecology. These in turn can inform vehicle and fuel infrastructure design and policymaking to help ensure we achieve the desired benefits.

We note that mobile electricity in the form of integrated mobility-electricity-information platforms places similar pressures on all automotive energy pathways. Thus, our discussion of a future in which society attempts to invoke a new sociotechnical system to support continued democratization of mobility, energy, and communications shifts the frame of reference. FCVs are not to be assessed only relative to today's ICEVs or HEVs. Rather, a comparison with ICEVs and HEVs that are themselves redesigned to be integrated mobility-electricity-information platforms is also required.

Conclusion: A Future for Hydrogen and FCVs?

We have argued that automobiles will be transformed from primarily mobility tools into mobile activity locales. Technologically, this will be facilitated by the integration of three of the supporting infrastructures built by modern societies during the twentieth century—automobility, electricity, and information systems. Such integration is in synchrony with the processes of societies becoming and being modern, in particular the central role of mobility in a posttraditional world that is dominated by lifestyle goals. It is in synchrony with continuing trends in the spread of mobile electricity and mobile communications into new services and across the population.

The deployment of automobility, electricity, and information infrastructures over the past 125 years was driven by mutually supporting relationships—positive feedback—between social forces and technological capabilities. Mobility begat more mobility; energy use begat more energy use; communications begat more communications. Initial evidence suggests that their integration is also driven by positive feedback—mobile communications begets more mobile communications. For those hoping communication might substitute for travel, the scholarship is at best ambiguous; many studies indicate that more communications leads to more, not less, travel. These relationships point to increased travel, energy use, and communications. They point to more mobility of people, commerce, and information in the future. If hydrogen and FCVs are to contribute to the collective goals of lowering criteria pollutant emissions, greenhouse gas emissions, and petroleum consumption, they will have to do so in this future.

Are FCVs the best energy system for integrated mobility-electricity-information platforms? In effect, are FCVs a better way to the future? If they are, they may gain competitive advantage as they tap into the long-term social and technological trends we have discussed in this paper. Ultimately, there is the potential of a renewable and direct solar, noncarbon energy path. In that potential lies the possibility for hydrogen, FCVs, and fuel cells more generally to capture the collective benefits.

A future of increased mobility, energy use, and information flow is not inevitable. However, we believe it is strongly compelled. A hydrogen energy system, made real through fuel cells, holds the potential to create this future while we meet goals to clean the air, protect the functioning of global ecosystems, and limit our dependence on domestic and imported petroleum. But the path to a future in which we achieve all this is neither certain nor obvious.

Acknowledgment

This work was supported by funding from the University of California Transportation Center, the University of California Energy Institute, and Toyota Motors Sales, USA Inc.

References

American Association of State Highway Officials. 1953. *Public Roads of the Past: Historic American Highways*. Washington, DC: American Association of State Highway Officials.

American Automobile Manufacturers Association. 1995. *Motor Vehicle Facts & Figures*. Washington, DC: American Automobile Manufacturers Association.

Cellular Telephone and Internet Association. 2003. *Semi-Annual Wireless Industry Survey*. Washington, DC: Cellular Telephone and Internet Association.

Davis, S. C., and S. W. Diegel. 2002. *Transportation Energy Data Book*. 22nd ed. Washington, DC: U.S. Department of Energy. http://www.osti.gov/bridge.

Federal Communications Commission. 2001. *Telephone Penetration by Income by State: Data through 2000*. Washington, DC: Federal Communications Commission, Industry Analysis Division, Common Carrier Bureau. http://www.fcc.gov/Bureaus/Common_Carrier/Reports/FCC-State_Link/IAD/pntris00.pdf.

———. 2003. *Trends in Telephone Service*. Washington, DC: Federal Communications Commission, Industry Analysis and Technology Division, Wireline Competition Bureau. http://www.fcc.gov/Bureaus/Common_Carrier/Reports/FCC-State_Link/IAD/trend803.pdf.

Giddens, A. 1984. *The Constitution of Society*. Berkeley, CA: University of California Press.

———. 1991. *Modernity and Self-Identity*. Stanford, CA: Stanford University Press.

Hägerstrand, T. 1970. What About People in Regional Science? *Papers of the Regional Science Association* 24: 7–21.

Harris Interactive. 2002. Two-thirds Hit the Net. Rochester: NY: Harris Interactive Inc. http://cyberatlas.internet.com/big_picture/geographics/article/0,,5911_1011491,00.html.

Hu, P. 2003. *Americans and Their Vehicles*. Oak Ridge, TN: Oak Ridge National Laboratory, Center for Transportation Analysis. July 22, 2003. http://nhts.ornl.gov/2001/presentations/americanVehicles/index.shtml.

Internet Software Consortium. 2002. Internet Domain Survey Host Count. Redwood City, CA: Internet Software Consortium. http://www.isc.org.

Jakle, J. A., and Keith A. Sculle. 1994. *The Gas Station in America*. Baltimore, MD: Johns Hopkins University Press.

Kurani, K. S., and T. Turrentine. 2002a. *Marketing Clean and Efficient Vehicles: A Review of Social Marketing and Social Science Approaches*. Davis, CA: Institute of Transportation Studies. UCD-ITS-RR-02-01.

Kurani, K. S., and T. Turrentine. 2002b. Household Adaptations to New Personal Transport Options: Constraints and Opportunities in Household Activity Space. In *In Perpetual Motion: Travel Behavior Research Opportunities and Application Challenges*. ed. H.S. Mahmassani. New York: Pergamon Press.

Kurani, K. S., T. Turrentine, and D. Sperling. 1994. Demand for Electric Vehicles in Hybrid Households: An Exploratory Analysis. *Transport Policy* 1 (October): 244–256.

———. 1996. Testing Electric Vehicle Demand in "Hybrid Households" Using a Reflexive Survey. *Transportation Research* D. 1(2): 131–150.

Lutzenhiser, L. 1995. Social and Behavioral Aspects of Energy Use. *Annual Review of Energy and Environment* 118: 247–289.

McShane, C. 1994. *Down the Asphalt Path: The Automobile and the American City*. New York: Columbia University Press.

New Deal Network. 2003. *TVA: Electricity for All: Rural Electrification.* New York: Franklin and Eleanor Roosevelt Institute, Teachers College/Columbia University. http://newdeal.feri.org/tva/tva10.htm.

Rosa, E. A., G. E. Machlis, and K. M. Keating. 1988. Energy and Society. *Annual Review of Sociology* 14: 149–172.

Sheller, M., and J. Urry. 2000. The City and the Car. *International Journal of Urban and Regional Research* 24(4): 737–757.

Stern, P. C. 1992. What Psychology Knows about Energy Conservation. *American Psychologist* 47(10): 1124–1132.

Stilgoe, J. R. 2001. *Roads, Highways, and Ecosystems.* Research Triangle Park, NC: National Humanities Center. http://www.nhc.rtp.nc.us/tserve/nattrans/ntueseland/essays/roads.htm.

Tedeschi, B. 2003. Eating Out and Logging On. *New York Times,* June 14.

Turrentine, T., and K. S. Kurani. 1998. Adapting Interactive Stated Response Techniques to a Self-completion Survey. *Transportation* 25(2): 207–222.

U.S. Department of Agriculture. 1985. *A Brief History of the Rural Electric and Telephone Programs.* Washington, DC: U.S. Department of Agriculture, Rural Electrification Administration.

U.S. Department of Commerce. 2002a. *American Housing Survey for the United States: 2001.* Washington, DC: U.S. Department of Commerce, Bureau of the Census. http://www.census.gov/hhes/www/housing/ahs/ahs01-80/tab25.html.

———. 2002b. *Statistical Abstract of the United States—2001.* Washington, DC: U.S. Department of Commerce, Bureau of the Census. http://www.census.gov/prod/www/statistical-abstract-02.html.

U.S. Department of Energy. 2002. *Progress Report for Hydrogen, Fuel Cells, and Infrastructure Technologies Program: Fiscal Year 2002 Progress Report.* Washington, DC: U.S. Department of Energy.

———. 2003. *National Petroleum News Survey.* Washington, DC: U.S. Department of Energy, Available at: http://www.ott.doe.gov/facts/archives/fotw279.shtml.

U.S. Department of Transportation. 1976. *America's Highways 1776/1976: A History of the Federal-Aid Program.* Washington, DC: U.S. Department of Transportation, Federal Highway Administration.

———. 1996a. *Highway Statistics Summary to 1995.* Washington, DC: U.S. Department of Transportation, Federal Highway Administration. Office of Highway Information Management. http://wwwcf.fhwa.dot.gov/ohim/summary95/index.html.

———. 1996b to 2001. *Highway Statistics.* Washington, DC: U.S. Department of Transportation, Federal Highway Administration. Office of Highway Information Management, annual. http://wwwcf.fhwa.dot.gov/policy/ohpi/hss/hsspubs.htm.

———. 2001. *Highway Statistics, 2001.* Washington, DC: U.S. Department of Transportation, Federal Highway Administration. Specific table available at http://wwwcf.fhwa.dot.gov/ohim/hs01/hm12.htm.

———. 2003. *NHTS 2001 Highlights: Report BTS03-05.* Washington, DC: U.S. Department of Transportation, Bureau of Transportation Statistics. http://www.bts.gov/publications/national_household_travel_survey/highlights_of_the_2001_national_household_travel_survey/index.html?submit=view+Online.

Winkle, T. A. 2002. Cellular Tower Proliferation in the United States. *The Geographical Review* 92 (January): 45–62.

CHAPTER 4

Fuel Cell Hybrid Vehicles: The Challenge for the Future

Taiyo Kawai

Toyota Motor Corporation has faced and overcome many challenges in its effort to create the ultimate eco car for the twenty-first century. In 1997, based on its accumulated experience and expertise, Toyota introduced the Prius, the world's first mass-produced hybrid electric vehicle, as an important step in this direction. The importance of hybrid electric technology to Toyota's vision of the ultimate eco car is shown in Fig. 4-1. Welcomed as an environmentally friendly car around the globe, the Prius achieves the highest fuel efficiency of any five-passenger mass-produced gasoline vehicle. Toyota considers the hybrid technology it developed through the creation of the Prius to be a core technology for the eco cars of tomorrow. The company continues to expand the use of its hybrid electric drivetrain across its product lineup.

There are several environmental issues facing automobiles today: cleaner exhaust, reduced carbon dioxide (CO_2) emissions, and reduced use of fossil fuels to ensure future energy security. The Prius is just one of many eco cars to meet these challenges.

The next step on the road to the ultimate eco car after hybrid electric vehicles is to use fuel cells in place of the internal combustion engine. Since setting out in 1992 to create fuel cell vehicles, Toyota has made a point of keeping fuel cell development in-house. The fuel cell is no exception to Toyota's basic stance of developing core technologies on its own. Compared to their application in the Gemini space missions in the 1960s, fuel cells have grown astonishingly small and powerful, evolving rapidly since vehicular applications began in earnest.

Toyota's completely in-house fuel cell—the Toyota FC Stack—not only has applications in its own fuel cell vehicle, but also in a variety of fuel cell products developed in cooperation with other industry-related companies.

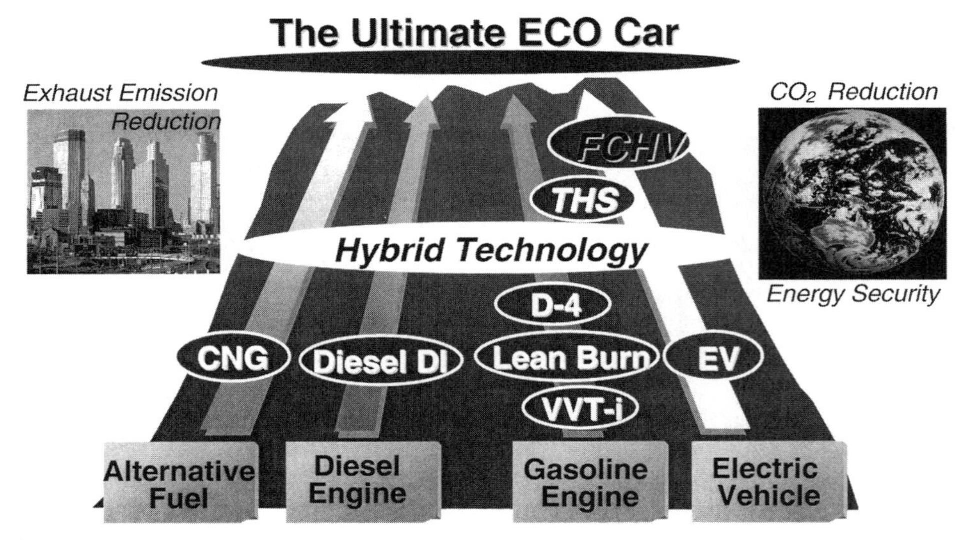

FIGURE 4-1. Toyota's conceptual pathways for the ultimate eco car.

Currently, Toyota has three types of fuel cell hybrid vehicles (FCHVs), differentiated by fuel system: compressed hydrogen storage, metal hydride hydrogen storage, and gasoline reformation. The Toyota FC stack is integrated into a light vehicle developed jointly with Daihatsu and a large bus built in cooperation with Hino. The FC stack is also being used in non-automotive applications. A 1-kilowatt (kW) cogeneration system is also being developed with the Aisin Seiki. The range of fuel cell vehicles built by Toyota is shown in Fig. 4-2.

Overall, the FC stack is a performance leader among vehicular fuel cells worldwide and is already on the verge of surpassing gasoline engines in power density. Figure 4-3 displays the Toyota FCHV that was recently released in limited number in the United States and Japan. It is a spacious, comfortable, and fun to drive vehicle, based on the Toyota Highlander, a five-passenger midsize sport utility vehicle.

The FCHV in Fig. 4-3 stores hydrogen in four 35 megapascal (MPa), equal to 5000 pounds per square inch (psi), high-pressure tanks. The system combines hydrogen from the tanks with oxygen from the air to generate 90 kW of electric power. The maximum speed of the Toyota FCHV is 155 kilometers per hour, and the range is 300 kilometers (km) measured by the Japanese 10-15 test mode, or somewhat less if tested on standard U.S. test cycles.

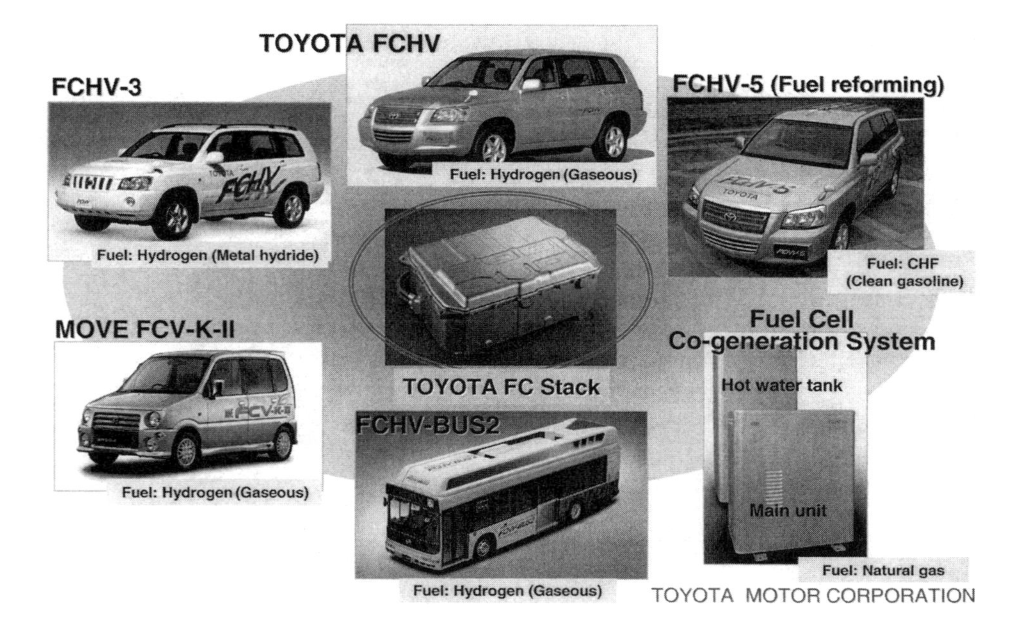

FIGURE 4-2. Toyota fuel cell technology.

Vehicle	Length/Width/Height	4,735/1,815/1,685mm	Fuel	Type	Pure hydrogen
	Maximum speed	155km/h		Storage method	High Pressure Tank
	Range (Japanese10.15mode)	300km		Maximum storage pressure	35 MPa
	Maximum number of passengers	5 passengers	Price	30 months lease contract	$10K/month

TOYOTA MOTOR CORPORATION

FIGURE 4-3. The Toyota FCHV fuel cell vehicle.

Toyota introduced the world's first fuel cell vehicles into the commercial market in 2002, simultaneously with Honda. As shown in Fig. 4-4, the first customers were the Japanese government, which leased four FCHVs, and the University of California at Irvine and the University of California at Davis, which leased one FCHV each. As of May 2003, these six FCHVs had logged over 17,000 km. User response has been positive. Customers appreciate the smooth acceleration and quiet, comfortable ride.

On the other hand, the range of the FCHV must be improved for it to become a true market-ready vehicle. Shorter start-up time (i.e., less than 5 seconds) is another area for improvement in the FCHV.

The hybrid system proven in Toyota's hybrid electric Prius achieves its highly efficient operation through sophisticated energy management of a gasoline engine and secondary battery. Toyota applied this hybrid technology to realize high efficiency in the FCHV as well (see Fig. 4-5).

The FCHV uses the same drivetrain as the Prius but the fuel cell stack is inserted in place of the gasoline engine. As in the Prius, a secondary battery provides additional power. The combination of a secondary battery with a fuel cell stack is more easily achieved than combining a secondary battery with a gasoline engine since the output for both sources in the FCHV is direct current electricity.

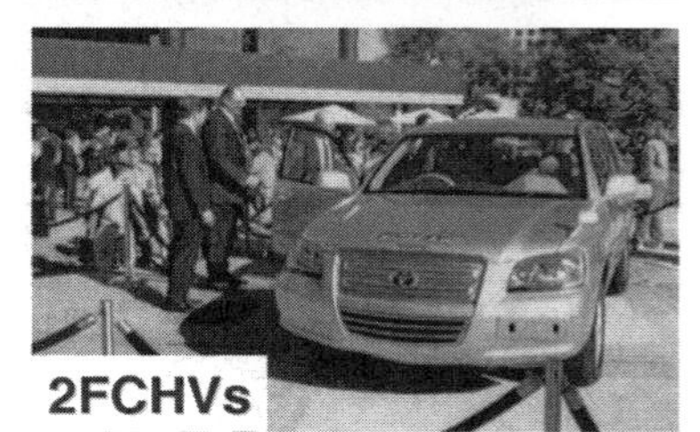

FIGURE 4-4. Toyota FCHVs in Japan and the United States.

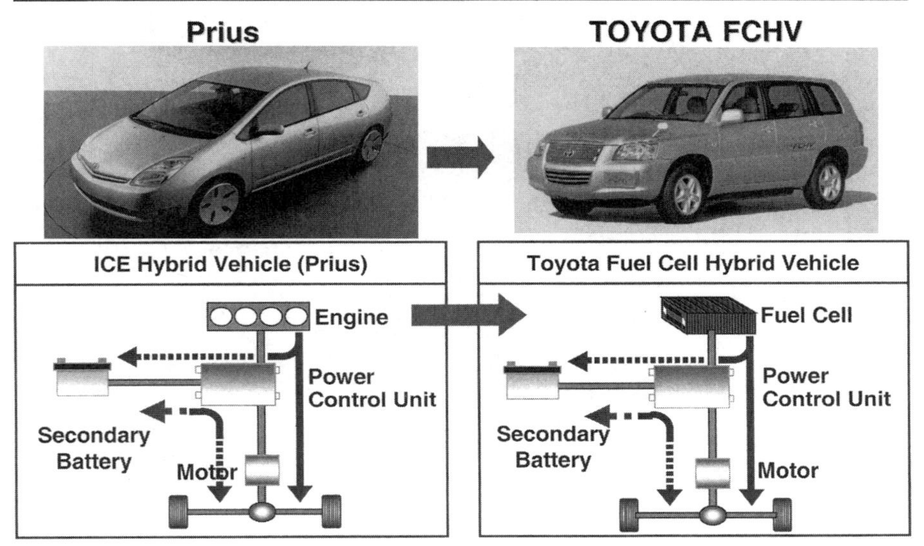

FIGURE 4-5. Comparison of Prius and FCHV.

By selecting the most efficient conversion system and using regenerative braking, Toyota is trying to accomplish maximum mileage efficiency. In the case of a fuel cell hybrid system, a secondary battery is the main power source when the vehicle is idling or traveling at low speeds. At normal driving speeds, the fuel cell is the main power source. During acceleration, electric power is supplied from both the fuel cell and the secondary battery. Through the regenerative braking system, the deceleration energy can be recovered by recharging the battery. This combination of a fuel cell and a secondary battery is so effective that Toyota is offering the system to other automakers.

Vehicle power needs fluctuate over a wide range during normal driving. A hybrid system is effective in achieving maximum efficiency. Figure 4-6 shows system efficiency versus output power for hybrid electric drive systems.

The conventional way to compare the efficiency of vehicles is by the distance they can travel per unit of fuel (e.g., kilometer per liter, miles per gallon). However, in Toyota's efforts to commercialize a fuel cell vehicle, two important factors must be considered: first, the vehicle's efficiency (tank-to-wheel), and secondly, how efficiently energy is obtained and transported to the vehicle's tank (well-to-tank). This new overall measure of efficiency, called "well-to-wheel," is important since Toyota's goals are to both reduce carbon dioxide emissions and consumption energy.

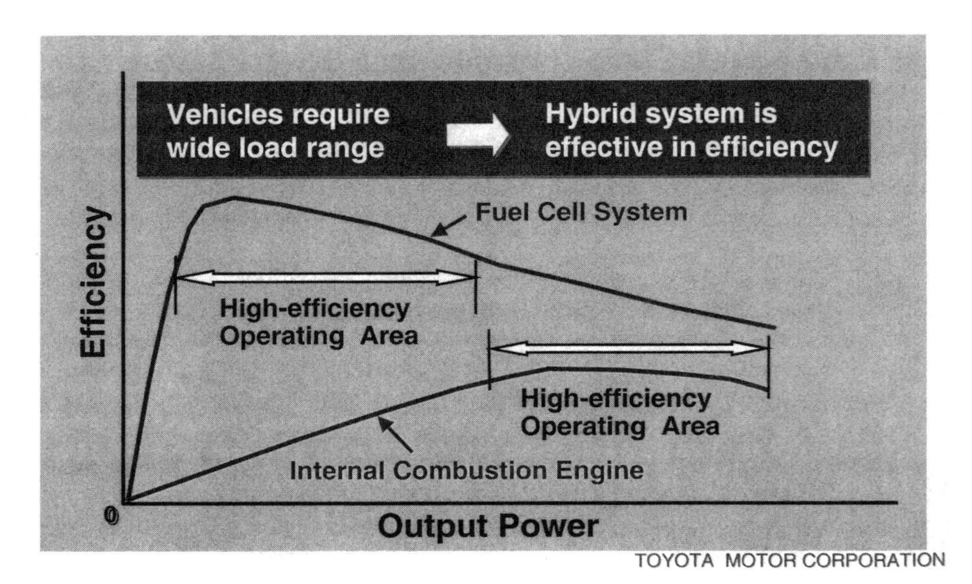

FIGURE 4-6. Hybrid system characteristics.

As indicated in Fig. 4-7, well-to-tank gasoline fuel efficiency is 88 percent and gasoline vehicle efficiency is 16 percent, resulting in a 14 percent well-to-wheel total efficiency. Using hybrid technology, the number is doubled to 28 percent.

For fuel cell hybrid vehicles, today's vehicle (tank-to-wheel) efficiency is 50 percent, but the well-to-tank efficiency is low. Toyota has set a target of 42 percent total well-to-wheel efficiency (which is three times higher than that of current gasoline vehicles). To achieve this goal, fuel cell vehicle efficiency must be over 60 percent. Simultaneously, Toyota has also requested the energy industry to develop high-efficiency hydrogen production methods to achieve 70 percent or more well-to-tank efficiency.

While the Toyota FCHV has the potential for high vehicle efficiency, several technical and environmental challenges remain, as shown in Fig. 4-8. First, vehicle performance and efficiency must be improved. Improvements in hydrogen storage must be achieved before the vehicle can be commercially marketed. Safety is another major concern. Further testing must be performed to determine and reduce the risk associated with carrying hydrogen and high voltage batteries. Finally, maintenance and disposal issues must be addressed. Costs of servicing the vehicle during its use, as well as the environmental impact from recycling and disposal of the vehicle at its end of life, require further investigation.

Wheel-to-Tank (%)
Tank-to-Wheel (%)
Well-to-Wheel (%)
0
10
20
30
40
Gasoline Vehicle
88
16
14%
28→32%
Prius (Gasoline HV)
88
32→37
Hydrogen FCV
FCHV
58*1
38*2
50*2
22%
29% With HV control
FCHV (Target)
70
60
3 × Gasoline, 1.5 × HV
42%
*1 Natural gas base
*2 Measurement from the electric current
Japanese 10-15 Mode Toyota's estimation
TOYOTA MOTOR CORPORATION

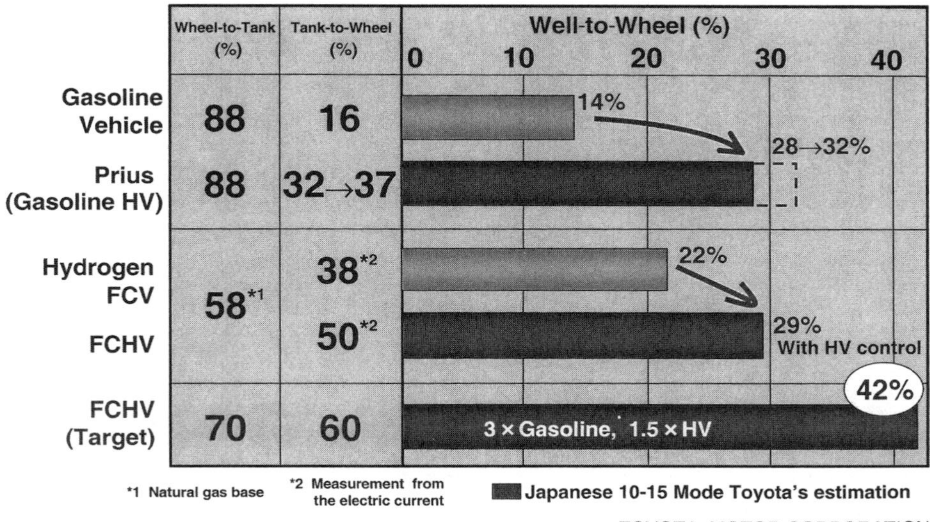

FIGURE 4-7. Well-to-Wheel efficiency.

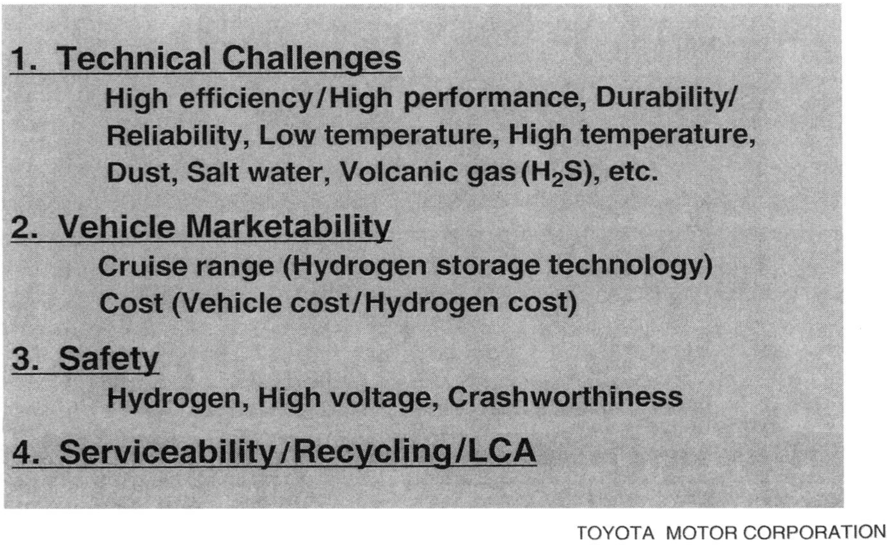

1. Technical Challenges
High efficiency/High performance, Durability/ Reliability, Low temperature, High temperature, Dust, Salt water, Volcanic gas (H_2S), etc.
2. Vehicle Marketability
Cruise range (Hydrogen storage technology)
Cost (Vehicle cost/Hydrogen cost)
3. Safety
Hydrogen, High voltage, Crashworthiness
4. Serviceability/Recycling/LCA
TOYOTA MOTOR CORPORATION

FIGURE 4-8. Issues for market introduction.

One of the more immediate challenges that Toyota is working on is low temperature operation. Low temperature start and performance is one of the main technical challenges facing the FCHV. In the FCHV, the fuel cell combines hydrogen and oxygen to form electricity, with pure water being the only emission, as shown in Fig. 4-9. However, this exhaust water freezes at low temperatures, making it impossible for the fuel cell to operate. Although this problem is simple to understand, it is one of the hardest obstacles to overcome. This is an example of a problem that is not observed in the laboratory but is a huge problem in real world applications.

As discussed earlier, the FCHV's limited driving range is one of the largest hurdles to its marketability. Hydrogen can be stored as a compressed gas, in an absorbing alloy, or in liquid form. Storage options are shown in Fig. 4-10. There are several issues associated with each storage method. To that end, advanced technologies such as carbon nanotubes and chemical hydrides are also being researched. However, these are still in the preliminary research stage and not ready for application in actual vehicles.

Overall, fuel cell technology is still underdeveloped. Many technical challenges remain and much more research needs to be done in order to achieve technical breakthroughs. Two things need to improve the technical level of fuel cell vehicles to that of existing internal combustion engines and gasoline-electric hybrids. First, at the basic research level, there needs

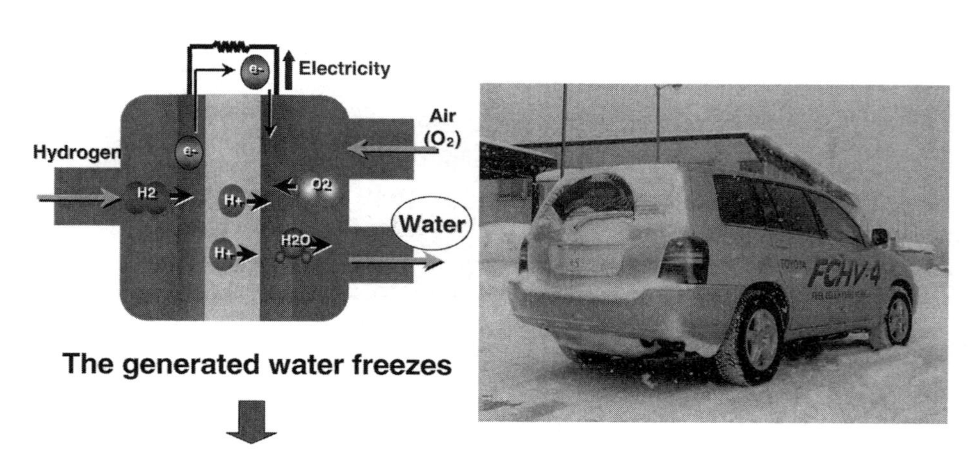

FIGURE 4-9. Low temperature performance.

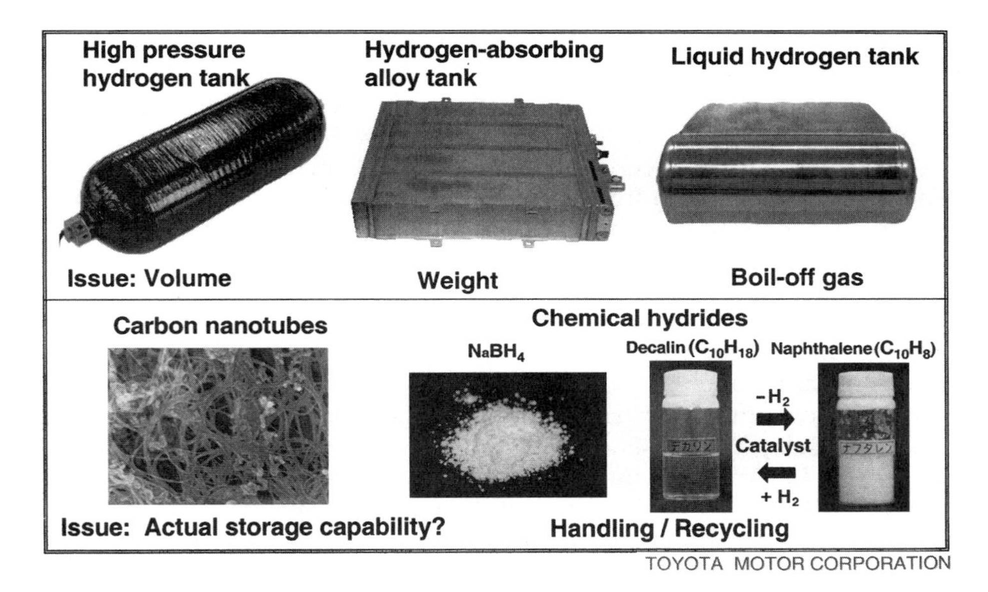

FIGURE 4-10. Hydrogen storage technology.

to be a formulation of international cooperation and framework for exchanging research outcomes. There are too many questions for one company or even industry to take on. An organized body or structure would facilitate information sharing. Secondly, competition among automotive manufacturers will help stimulate production improvements and commercialization. This allows automotive manufacturers to pursue various strategies to make fuel cell vehicles successful at a retail level.

Overcoming Barriers to Commercialization

Toyota believes there are three essential conditions for successful dissemination of fuel cell vehicles. First, there is the issue of vehicle marketability. Technical challenges discussed earlier must be resolved, and the cost of the vehicle must be at a level that the market can support. As with other vehicles today, fuel cell vehicles must be packaged in a way that is attractive to the consumer. Also, there need to be plans in place that anticipate a hydrogen society. Finally, there must be a consumer shift to understand and prefer environmental vehicles.

To promote a smooth shift to a hydrogen-based society, we need the combined effort of all sectors of society, including government. This is clearly the case when it comes to developing hydrogen manufacturing and storage technology, as well as building the hydrogen fueling station infrastructure. We must invest in training new specialists in universities and technical schools. And codes and standards must be established in a comprehensive and multifaceted way by society.

Hydrogen is an energy carrier that can be made from a variety of sources. What it is made from and how it is made determine the potential quantity of supply, as well as the well-to-tank efficiency and amount of CO_2 produced. Production cost must also be taken into account. To create a more affluent future society through the use of hydrogen energy, we must think carefully and thoroughly about how best to make and transport hydrogen, build infrastructure such as hydrogen fueling stations, and promote technological advances.

The most difficult issue for market introduction is the hydrogen infrastructure. Today, there is practically no hydrogen infrastructure. As shown in Fig. 4-11, the hydrogen infrastructure will be quite complex. However, as hydrogen can be produced from a variety of energy sources, there needs to be an international long-term energy policy to promote development. Development of a cheap hydrogen supply chain is absolutely necessary for expanding the use of fuel cell vehicles.

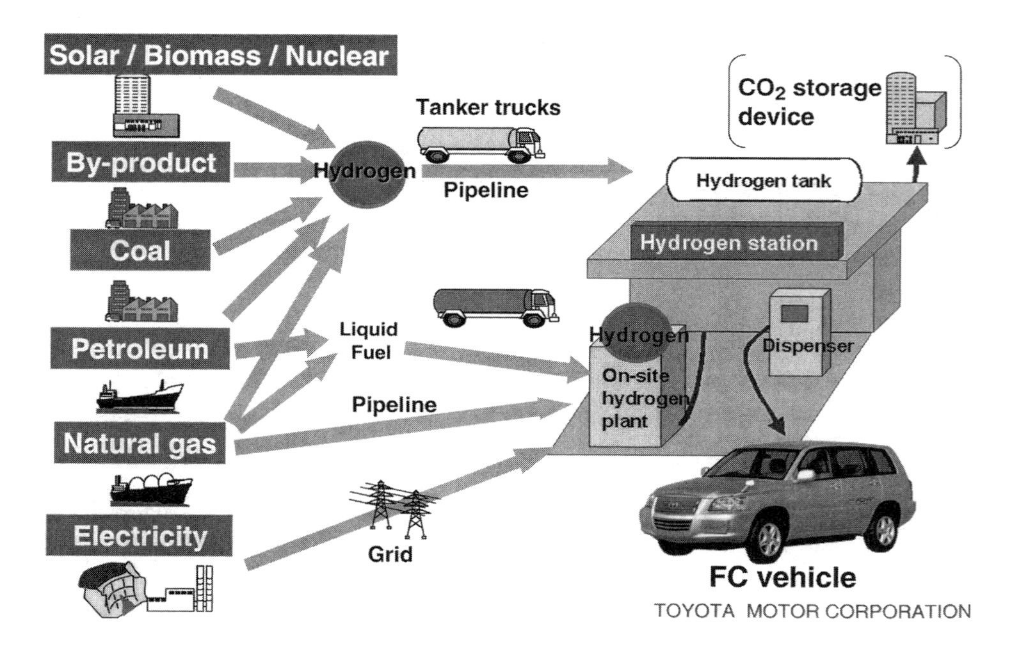

FIGURE 4-11. Schematic of hydrogen infrastructure pathways.

Figure 4-12 shows the well-to-wheel CO_2 emission rates for a variety of vehicle types. Although a hydrogen-powered fuel cell vehicle puts out no CO_2 itself, the plants that make the hydrogen do discharge CO_2. So if we intend to reduce CO_2 by popularizing fuel cell vehicles, we must go beyond tank-to-wheel efficiency and also tackle well-to-tank efficiency. For example, coal is an abundant and economical natural resource, but making hydrogen from coal produces large amounts of CO_2. And while it is true that we can make hydrogen from biomass or use "green energy" such as solar power to produce hydrogen from water through electrolysis, these methods cannot yet supply more than a limited quantity. We need to develop a "best mix strategy" of methods for making and transporting hydrogen, while considering all the factors involved, including CO_2 output, cost, and supply volume requirements.

The Japan Hydrogen and Fuel Cell Demonstration Project (JHFC), summarized in Fig. 4-13, is a program set up by the Ministry of Economy, Trade and Industry (METI) in March 2003. In this program, hydrogen refueling facilities with different types of fuel and production methods will be tested. Vehicles from five automakers are taking part in this project, including Toyota's FCHV and fuel cell bus.

Mass production of fuel cell vehicles will start after the total cost of the fuel cell vehicle becomes lower than that of a gasoline vehicle. As shown

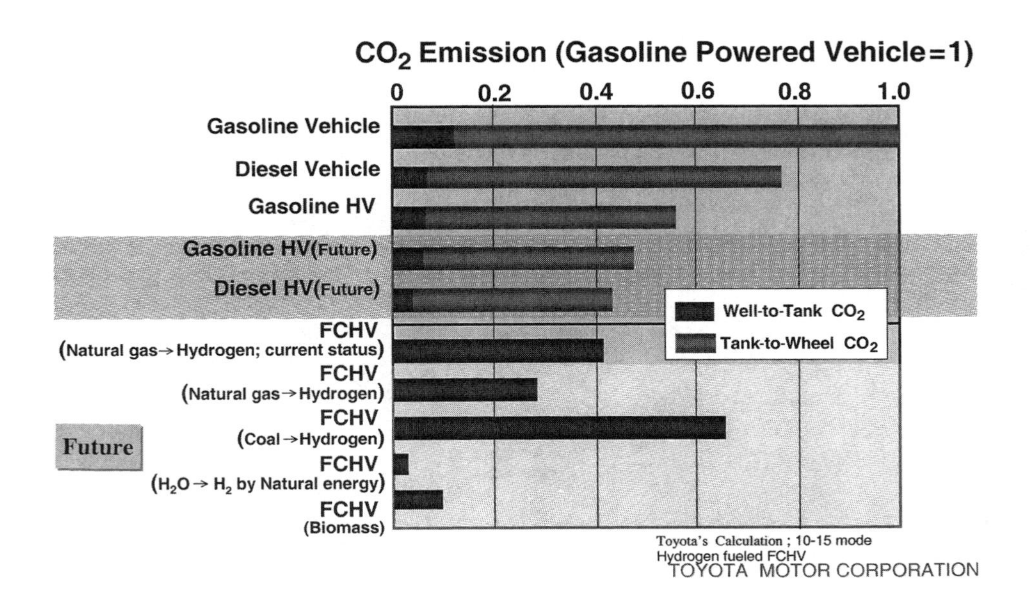

FIGURE 4-12. Well-to-Wheel carbon dioxide emissions.

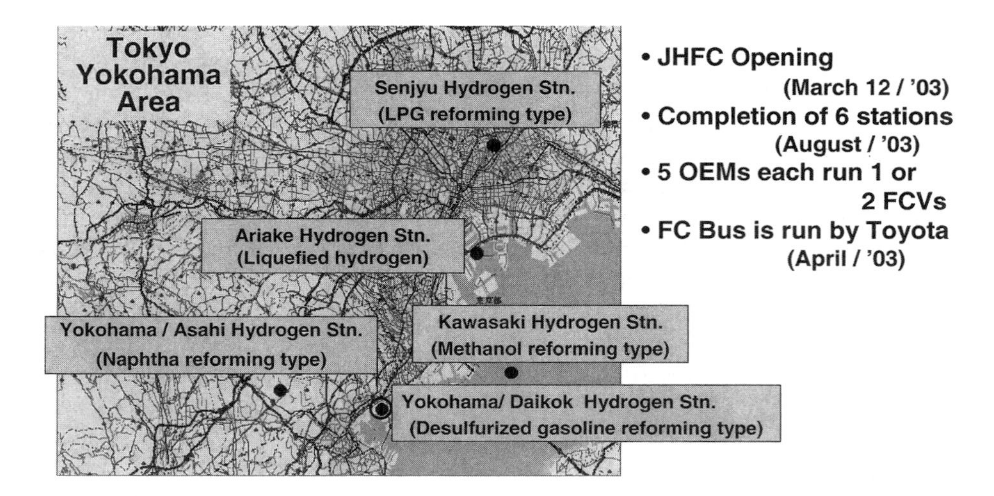

FIGURE 4-13. The Japan hydrogen and fuel cell demonstration project.

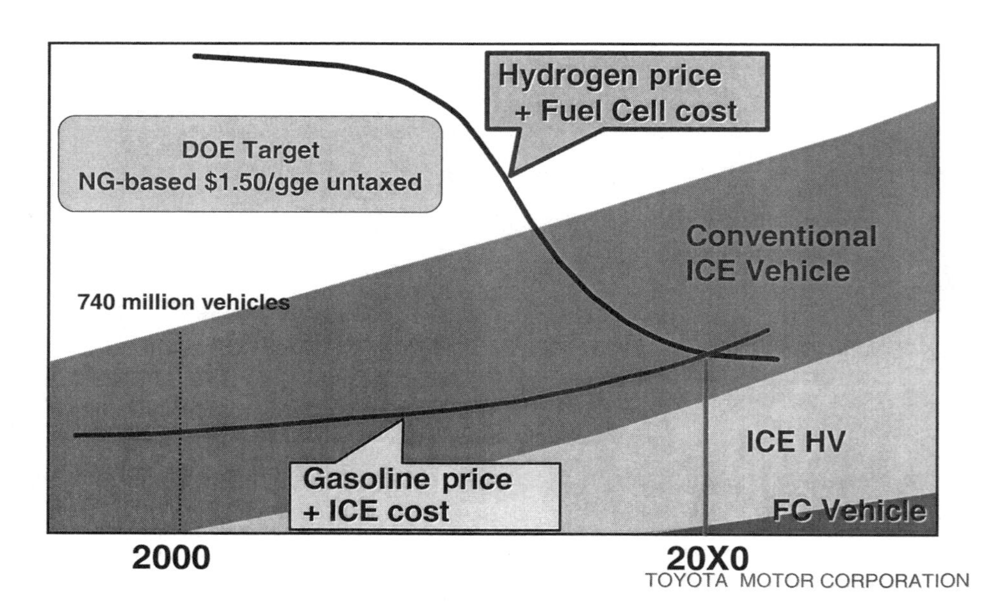

FIGURE 4-14. Hydrogen production costs.

in Fig. 4-14, the U.S. Department of Energy has set a target cost of hydrogen that is close to the retail gasoline price in the United States. When that happens, customer's cost for the hydrogen fuel for fuel cell vehicles will be less than one-third compared to that of a gasoline vehicle because of the high efficiency of the fuel cell vehicle. The realized savings from fuel cell vehicles must be a strong driving force for the hydrogen economy. We anticipate this can happen around the year 2020 or 2030.

There are many milestones to reach in order to get to the point of mass production of fuel cell vehicles within 10 years. For widespread introduction of fuel cell vehicles, we must build a global network among engineers and scientists that allows not only for competition but also for collaboration, as shown in Fig. 4-15.

To build a hydrogen society, every sector should establish a target and collaborate more closely. If we succeed in the creation of these new values and resolve global environmental and energy security issues, we can lead future generations into a new era of transportation.

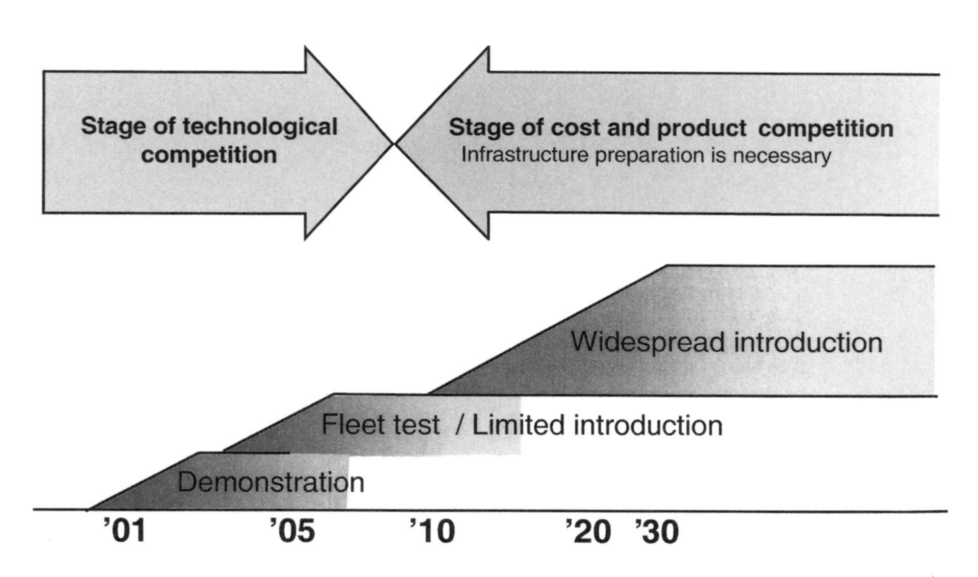

FIGURE 4-15. Forecast of fuel cell vehicle market introduction.

CHAPTER 5

Where Will the Hydrogen Come From? System Considerations and Hydrogen Supply

Joan Ogden

Much of the current interest in hydrogen stems from its potential to radically reduce several important societal impacts of transportation fuel use. Hydrogen can be made from renewable resources, fossil fuels decarbonized by the capture and secure storage of carbon dioxide (CO_2), or nuclear energy. Any one of these paths has the potential to produce hydrogen on a global scale with near-zero full fuel cycle emissions of greenhouse gases and greatly reduced emissions of air pollutants, although each faces challenges. Because it can be made from a variety of widely available primary sources, hydrogen could facilitate a diversification of the primary energy supply for transportation away from the present near-exclusive dependence on oil. Moreover, there is the intriguing possibility that hydrogen and fuel cells might enable improved vehicle designs and new features, such as onboard electricity generation, that would make them attractive to auto manufacturers and consumers, even without policies reflecting external costs of energy (Kurani *et al.*, see Chapter 3; Burns *et al.*, 2002).

Today hydrogen is produced almost exclusively from fossil fuels for chemical applications and oil refining, accounting for perhaps 1 percent of global primary energy use. Although commercial hydrogen production, storage, and distribution systems exist, and hydrogen end-use technologies are evolving rapidly, much work remains to be done to implement the long term vision of a zero-emission hydrogen transportation fuel supply. The degree to which hydrogen's potential benefits can be realized and the cost of doing so depend critically on how the hydrogen is produced and delivered, as well as how it is used. Clearly, not all "well-to-wheels" pathways for hydrogen production and use are equally attractive in terms of economics, resource availability, sustainability, security, or environmental advantages.

This chapter discusses how the evolution of a future hydrogen supply might be affected by the complex factors, and sometimes conflicting goals, that determine the best production and distribution pathways for a given location and level of demand. We take a systems perspective, focusing on relationships among various aspects of a hydrogen system—supply, demand, economics, externalities, technological progress—and what this implies about the lowest cost or lowest emission hydrogen supply over time.

In the next section, we describe several near term and long term options for hydrogen supply, and the factors that influence the design of a hydrogen energy system. In the four sections that follow, we address several key system-level questions for future hydrogen supply. First, what does the goal of minimizing externalities imply about hydrogen supply during the early stages of a transition when hydrogen demand is relatively small? And in the long term? Second, what does the goal of keeping infrastructure costs low during transition, by matching supply and demand and utilizing existing infrastructure, imply about which primary resources are used for hydrogen production? Third, what do regional and geographic factors imply about hydrogen supply in developed and developing countries? Finally, how might interactions with the existing infrastructure and rest of the energy system affect the choice of hydrogen supply?

Each question is explored in a separate section, organized as a series of observations, hypotheses, and further questions related to how hydrogen supply might evolve over time. Clearly, there are large uncertainties in trying to envision a transition to a future hydrogen energy system. This paper will not seek to give definitive answers but rather to stimulate discussion about how to develop the hydrogen supply to best realize hydrogen's potential benefits at an acceptable cost.

Future Hydrogen Supply Options

Because hydrogen can be produced from a variety of feedstocks, at a wide range of sizes from city scale to household level, and delivered in various ways ranging from truck delivery of cryogenic liquid to gas pipeline delivery of compressed gas, there are many possible supply options. (To simplify the analysis, all supply options in this chapter assume delivery to motor vehicles of hydrogen as a compressed gas.) Near term supply options include:

- Central steam reforming of natural gas with distribution of hydrogen via compressed gas or liquid hydrogen truck or pipeline
- Recovery of hydrogen from chemical processes with distribution of hydrogen

- Onsite production of hydrogen via small scale steam reforming of natural gas at the refueling station
- Onsite production of hydrogen via small scale water electrolysis at the refueling station

These near term options are shown schematically in Fig. 5-1. They can all be realized with commercially available technology and are used to

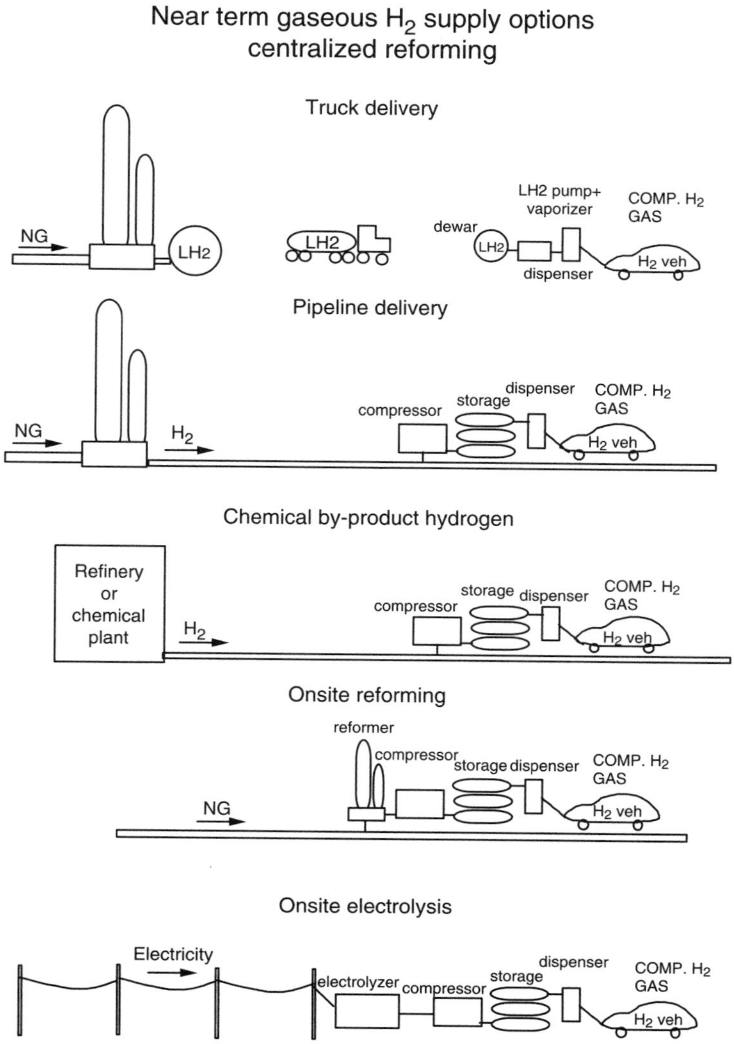

FIGURE 5-1. Near term gaseous hydrogen supply options.

produce hydrogen for chemical applications today, although smallscale onsite production systems are still undergoing rapid development for refueling station applications.

Long term hydrogen supply options include:

- Centralized production of hydrogen via electrolysis using CO_2-free electricity sources with distribution of hydrogen
- Decentralized solar or wind powered electrolysis
- Gasification of coal, petroleum coke, biomass, or wastes with capture and secure storage of CO_2
- Thermochemical water splitting powered by high temperature nuclear or solar heat

The long term options are shown schematically in Fig. 5-2. All these systems produce hydrogen with near-zero to zero emissions of greenhouse gases and air pollutants.

Challenges Facing Long Term Supply Options

Each of these near-zero emission, long term hydrogen supply options faces significant challenges before it could be implemented on a global scale.

For hydrogen from renewables, the issue is primarily cost. Electrolyzers using solar or wind power and biomass gasification systems could be built today using commercial or near-commercial technology, but hydrogen costs would be several times higher than for the near term supply options shown in Fig. 5-2. Although electricity from widely available, vast renewable resources like wind and especially solar is still much too costly for hydrogen production, in some locations off-peak hydropower or geothermal power could be cost effective for electrolytic hydrogen production. For biomass hydrogen, the limiting factors might be land use and competing uses for low cost biomass feedstocks in the electricity sector. There are a number of promising experimental renewable production methods such as photoelectrochemical and photobiological systems that are still in the basic science stage, but they have not been considered here.

Nuclear electrolytic hydrogen suffers from high cost as well, unless low cost off-peak power from a nuclear plant is used. Thermochemical water splitting systems powered by nuclear heat have been projected to offer lower hydrogen costs. However, these systems are still in the laboratory stage, face a number of technical issues, and should be regarded as less technically mature than gasification-based systems or electrolyzers. In addition there are issues of proliferation and waste disposal associated with nuclear energy.

Fossil hydrogen with CO_2 capture and sequestration holds the promise of near-zero emissions and a relatively low hydrogen production cost, assuming that nearby suitable CO_2 disposal sites are available and that

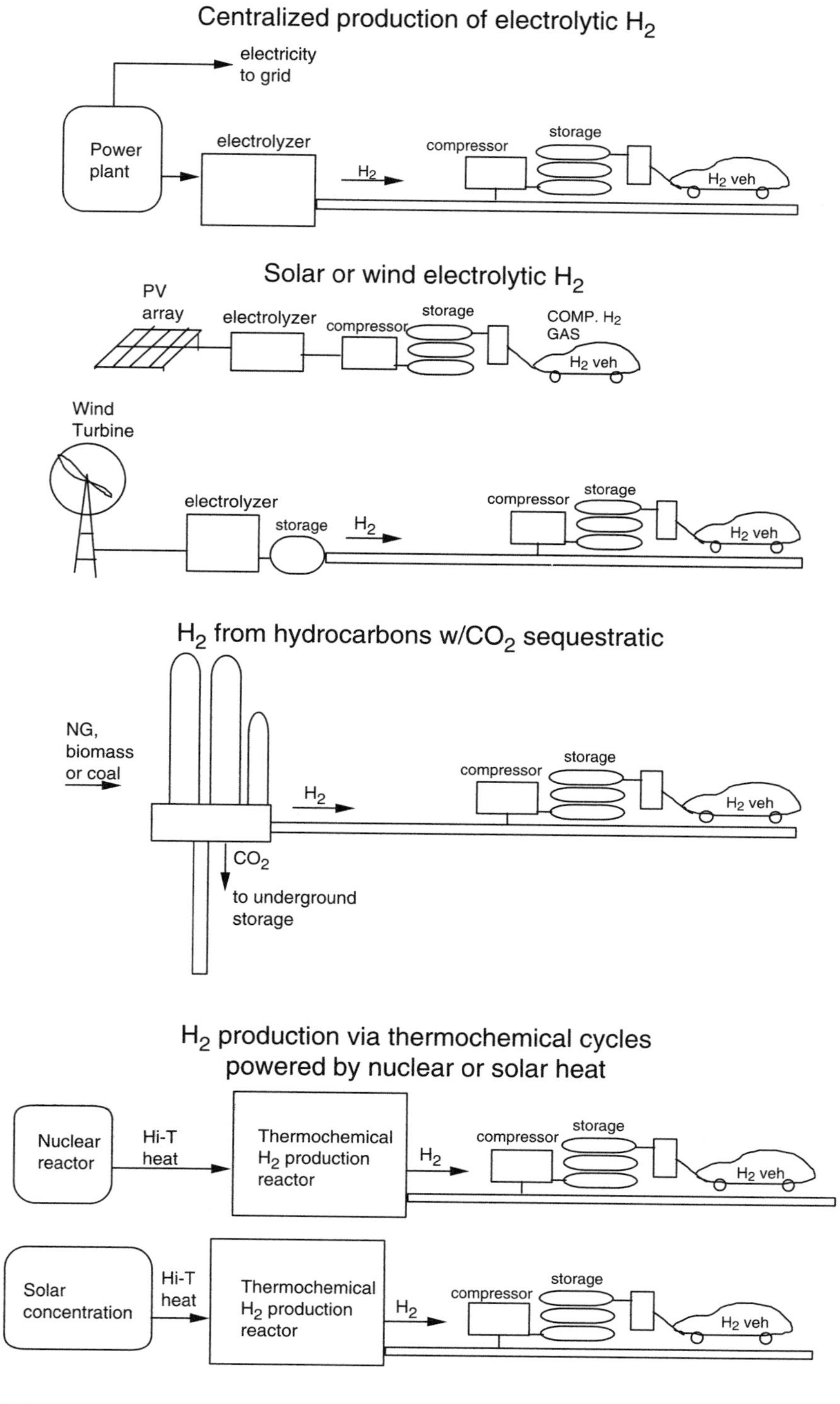

Centralized production of electrolytic H_2

electricity to grid

Power plant

electrolyzer

H_2

compressor

storage

H_2 veh

Solar or wind electrolytic H_2

PV array

electrolyzer

compressor

storage

COMP. H_2 GAS

H_2 veh

Wind Turbine

electrolyzer

storage

H_2

compressor

storage

H_2 veh

H_2 from hydrocarbons w/CO_2 sequestratic

NG, biomass or coal

H_2

compressor

storage

H_2 veh

CO_2

to underground storage

H_2 production via thermochemical cycles powered by nuclear or solar heat

Nuclear reactor

Hi-T heat

Thermochemical H_2 production reactor

H_2

compressor

storage

H_2 veh

Solar concentration

Hi-T heat

Thermochemical H_2 production reactor

H_2

compressor

storage

H_2 veh

FIGURE 5-2. Long term hydrogen supply options.

hydrogen is produced at large scale. (It is not economically feasible to collect CO_2 from small hydrogen production systems such as fueling stations or buildings with onsite re-formers.) Many of the technologies required for CO_2 sequestration already exist and have been used for enhanced oil recovery. Much remains unknown about the potential environmental impacts and feasibility of this concept (Williams, 2002). Several demonstration projects for CO_2 recovery are ongoing, with more planned, that should shed light on these issues (for example, the "FutureGen" program in the United States [U.S. Department of Energy, 2003a] and the IEA Greenhouse Gas Program in Europe are supporting large scale sequestration experiments [Gale and Kaya, 2003]).

Factors for Designing a Hydrogen Supply System

Before considering more complex questions about hydrogen system transitions, we discuss the factors that must be considered in designing a hydrogen supply system for a particular situation. Characteristics of the hydrogen demand are crucial to determining the best option for hydrogen supply. Important factors include the size of the demand; its geographic location and the geographic density of demand; the proximity to hydrogen supply, which determines the distribution technology; the daily and seasonal time variation of demand, which determines the amount of energy storage needed in the system; and the growth rate of the demand, which determines how fast a hydrogen energy supply system expands. The growth of hydrogen demand for vehicles will likely depend on progress in hydrogen technologies, such as onboard storage and fuel cells, and their competitiveness with other advanced vehicles.

Another major factor for designing a future hydrogen supply is the availability, location, abundance, and cost of primary energy resources, such as coal, biomass, or wind. Secondary energy carriers, such as natural gas or grid electricity, might be used for hydrogen production in regions where low cost natural gas or off-peak power is available. For hydrogen derived from fossil fuels, the availability and storage capacity of sites for carbon sequestration is also an issue.

Finally, technology and policy are "wildcards" that could greatly influence the design of future hydrogen supply. Many of the technologies that make up a hydrogen energy system are rapidly evolving. Advances in hydrogen technologies could help determine the choice of a hydrogen supply pathway. For example, the successful development of inexpensive small natural gas reformers or high pressure electrolyzers could encourage distributed hydrogen production at refueling stations. A breakthrough in hydrogen storage could change the way hydrogen is distributed to users. New vehicle types might create market pull to enable rapid growth of hydrogen demand. Policies reflecting the external costs of energy could encourage zero emission hydrogen supply technologies.

Key Questions for Future Hydrogen Supply

Having discussed possible hydrogen supply systems and general design criteria, we now pose a series of "system-level" questions about transition to a hydrogen energy system.

Minimizing Environmental Externalities in a Hydrogen Transition

Radical reduction of externalities is widely seen as the reason for considering hydrogen as an energy carrier. Clearly, the economic or societal value placed on reducing externalities associated with energy use through carbon taxes, air pollution regulations, or other strategies could be critical in determining the viability of any hydrogen transition in transportation. What does the goal of "zero emissions" say about hydrogen supply, especially during a transition when zero emission hydrogen supplies are likely to be expensive?

Depending on the hydrogen pathway, the well-to-wheels emissions of greenhouse gases and air pollutants can be less than or greater than those for advanced internal combustion engine vehicles fueled with gasoline or diesel (Thomas *et al.*, 1998a,b; Wang, 1999; General Motors *et al.*, 2001; Weiss *et al.*, 2000; Wang, 2002). Policy measures aimed at reducing externalities could become a powerful driver for low-emission hydrogen pathways. However, these policies alone might not be sufficient to encourage near term hydrogen options that offer only modest societal advantages over advanced vehicle technologies using conventional fuels unless they are seen as a necessary transitional stage toward a near-zero emission hydrogen system in the longer term. A key question is how far down the path toward zero emissions will society choose to go to meet environmental and security goals for the transportation sector. How cost-effective will hydrogen vehicles be as an environmental strategy versus, for example, extremely low emission, high-efficiency combustion vehicles?

Observations

Making hydrogen from fossil sources is currently the dominant method of production and is likely to remain dominant in the near term. As shown in Fig. 5-3, about 95 percent of hydrogen made on a large scale in 1999 was from fossil primary sources and about 48 percent of hydrogen globally was made from natural gas. In the United States, over 90 percent of hydrogen is currently derived from natural gas (U.S. Department of Energy, 2003b). In many parts of the world where low cost fossil resources are available, fossil-derived hydrogen is likely to remain the lowest cost option for several decades.

There could be modest societal benefits if hydrogen is made from fossil fuels in the near term, even without CO_2 sequestration. If hydrogen is derived from fossil sources without carbon sequestration and used in an

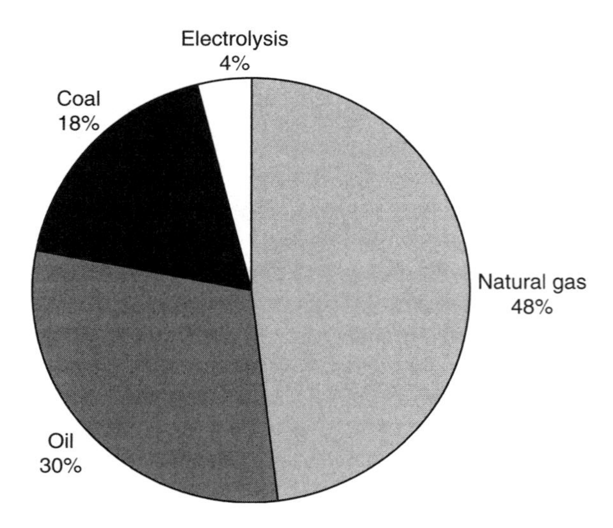

FIGURE 5-3. Global hydrogen production in 1999.

efficient hydrogen fuel cell or internal combustion engine hybrid vehicle, well-to-wheels greenhouse gas emissions are comparable to those for efficient diesel or compressed natural gas internal combustion engine hybrid vehicles and slightly less than those for gasoline hybrids (Thomas *et al.*, 1998; Weiss *et al.*, 2000; General Motors *et al.*, 2001; Wang, 2002; Williams, 2002). There is a much stronger greenhouse gas benefit to using fossil-derived hydrogen in vehicles as compared to advanced internal combustion engine vehicles using fossil-derived liquid fuels or compressed natural gas if CO_2 is captured and sequestered (see Fig. 5-4). In addition, there might be well-to-wheels air pollutant emissions benefits with fossil-derived hydrogen, even without CO_2 sequestration, because of lower direct emissions from the vehicle (Wang, 2002) (Fig. 5-5). Moreover, oil use for transportation would be reduced. Overall, external costs of energy would be reduced with near term use of hydrogen derived from natural gas, as compared to advanced vehicles using conventional liquid fuels (Ogden *et al.*, 2004).

There are a number of hydrogen supply pathways that offer near-zero well-to-wheels emissions of greenhouse gases and much reduced emissions of air pollutants. With the exception of hydrogen from locally important, but globally small, low-cost renewable resources, such as biomass, wastes, and off-peak hydro- or geothermal power, today's "near-zero" emission hydrogen supply systems are generally more costly than current fossil hydrogen systems. Zero emission electrolytic hydrogen supply options that could, in principle, be developed on a global scale, such as wind, solar, and nuclear energy, are currently several times as expensive as hydrogen from natural gas. Hydrogen production from fossil fuels with CO_2 capture and sequestration

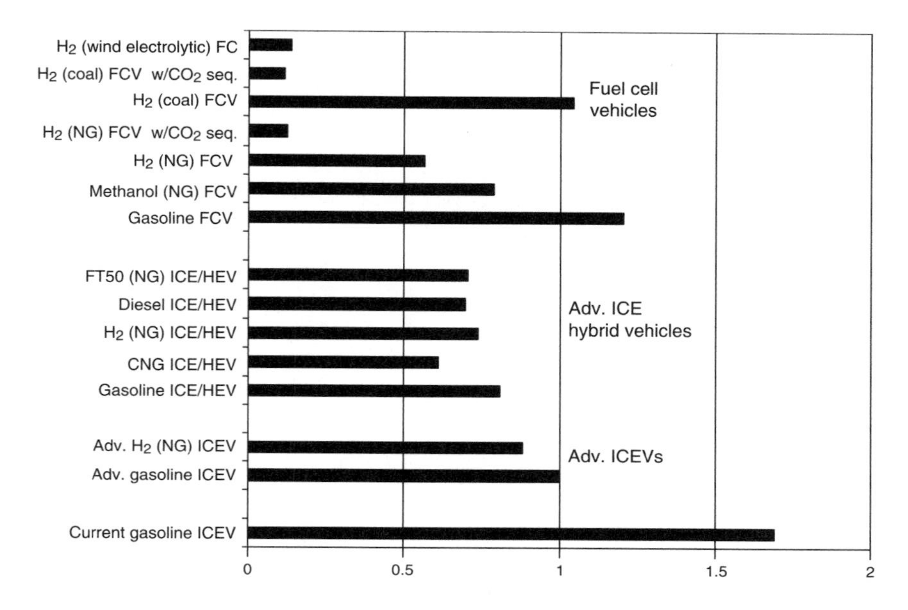

FIGURE 5-4. Well-to-wheels GHG emissions normalized to efficient gasoline vehicles.

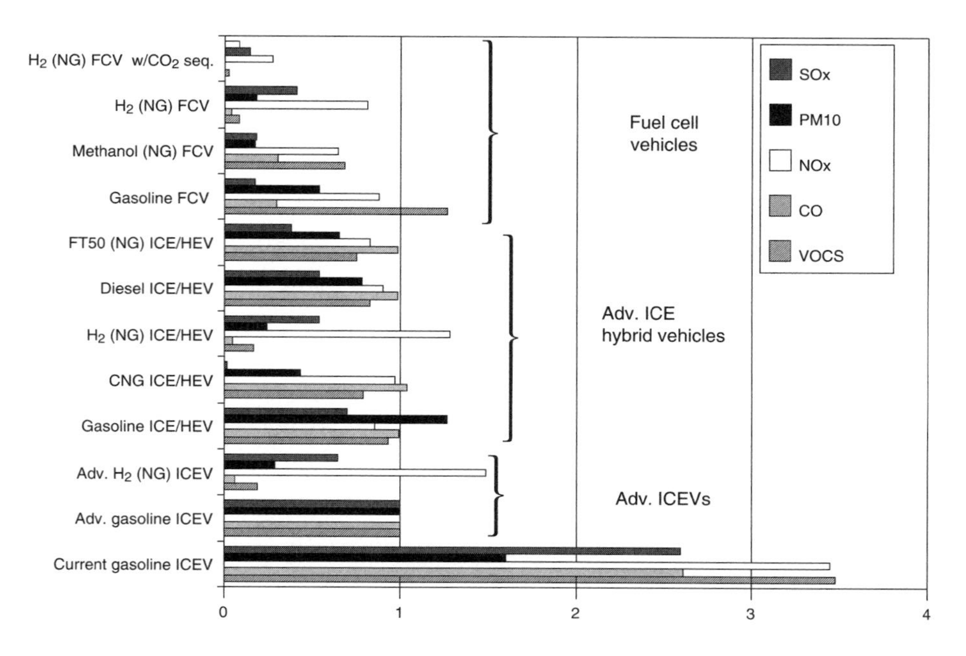

FIGURE 5-5. Well-to-wheels air pollutant emissions normalized to efficient gasoline vehicles.

holds the promise of simultaneously achieving low cost and low carbon emissions on a large scale, but issues remain about the viability of large scale sequestration of CO_2 (Gale and Kaya, 2003). Nuclear or solar thermochemical water splitting cycles have been proposed, but the technology is still in the research and development stage and involves technical, environmental, political, and scale issues (Brown *et al.*, 2002).

Stabilizing atmospheric CO_2 concentrations will ultimately require decarbonizing the fuel sector, starting within a few decades. Climate models suggest that to stabilize atmospheric concentrations of CO_2 in the range of 450 to 550 parts per million (ppm) (that might avoid serious, irreversible climate effect), it will be necessary to begin significantly decarbonizing the energy supply starting within a few decades (Wigley *et al.*, 1996; Nakicenovic *et al.*, 1998). As direct combustion of fuels for heating and transportation currently accounts for about two-thirds of global greenhouse gas emissions, it will be necessary to drastically reduce the carbon emissions from the fuel sector as well as the electricity sector. Even if the electricity sector were completely decarbonized by 2100, stabilization of CO_2 concentrations at 550 ppm would require a threefold decrease in emissions from direct use of fuels, including in transportation, as compared to a "business as usual" energy supply scenario (Williams, 2002).

Under rapid growth scenarios for hydrogen transportation fuel, it would take several decades before hydrogen vehicles would penetrate vehicle markets widely or use enough energy to make a large difference in reducing emissions or oil consumption. This is illustrated in Fig. 5-6, where projected hydrogen use is shown over time for four future hydrogen demand scenarios considered by the U.S. Department of Energy (Mintz and Singh, 2003). In a "business as usual" scenario developed by the Energy Information Administration, hydrogen plays a minor role, reaching only about 1 percent of the total light duty vehicle population by 2050 (Energy Information Agency, 2003). The other three scenarios include high levels of environmental awareness, rapid technical success, and concerted efforts to develop hydrogen vehicles, so that hydrogen captures 30–100 percent of light duty vehicle markets by 2050. According to the middle of these (based on the "2050 study" jointly carried out by the U.S. Department of Energy and National Resources of Canada), about 7 percent of the U.S. light duty vehicle fleet, or 20 million vehicles, would be powered by hydrogen in 2025. The total energy use in 2025 would be about 0.7 quadrillion BTUs (quads) of hydrogen per year. If this hydrogen were made from natural gas, about 1 quad or 3 percent of the projected natural gas supply in 2025 would be needed for hydrogen production.

Hypotheses

To fully realize hydrogen's multiple benefits in the long term, it will be important to use hydrogen supply pathways with near-zero emissions of

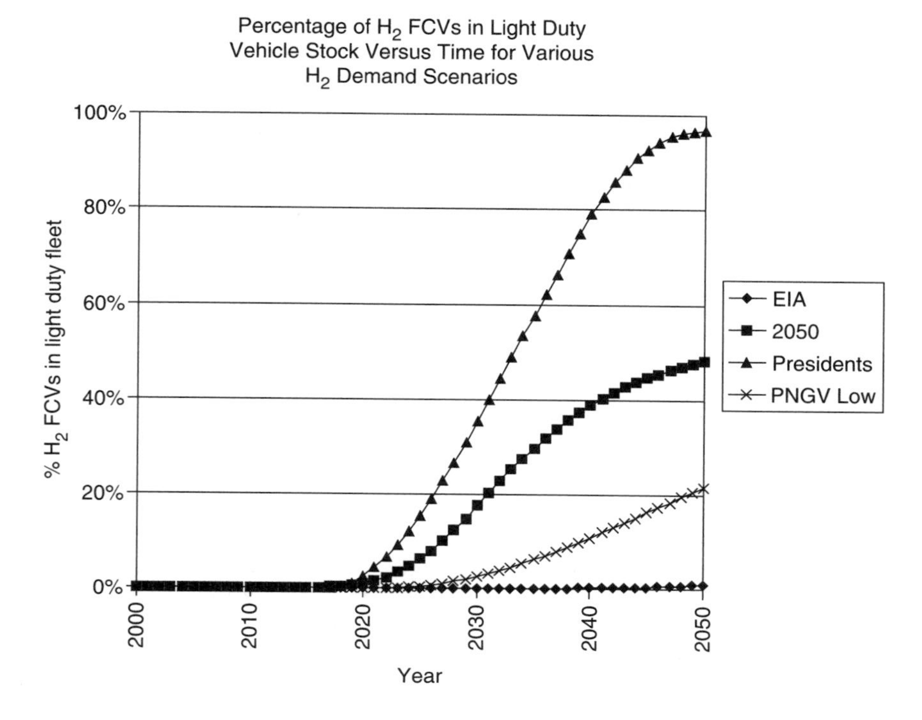

FIGURE 5-6. Scenarios for hydrogen demand for vehicles.

greenhouse gases and air pollutants, and to have the possibility of using diverse primary supplies. But in the near term, use of natural gas-derived hydrogen without carbon sequestration in efficient fuel cell or internal combustion engine cars would reduce air pollutant emissions and oil use per mile, although it would not significantly reduce greenhouse gas emissions per mile compared to advanced gasoline or diesel hybrids. The supply impact of natural gas use for hydrogen production would be modest. (Even the most aggressive demand scenario in Fig. 5-6 would require less than a 10 percent increase of U.S. natural gas use by 2025.) This suggests that natural gas could serve as a transitional source for near term production of hydrogen in the United States, with modest societal benefits compared to other advanced vehicles using conventional fuels, and a relatively small impact on natural gas use. Beyond this time frame, it will be important to begin a switch to near-zero carbon sources of hydrogen.

While developing hydrogen end-use technologies such as fuel cells, it is important to simultaneously develop near-zero emission hydrogen production technologies so that they could be widely employed in a few decades. Electrolysis from off-peak hydropower and geothermal power could be locally important, starting in the near term. Some important enabling technologies for zero-emission hydrogen might appear first in the electricity sector. For example, coal gasification with CO_2 sequestration and wind power might be in commercial use for electric generation in 20 to 30 years. Where low cost feedstocks are available, biomass hydrogen could also be important.

From the climate perspective, there might be a few decades to bring hydrogen vehicle technologies and near-zero greenhouse gas emission hydrogen production technologies to technical and commercial maturity before they would be needed on a massive scale to sharply reduce greenhouse gas emissions from the fuels sector. However, for hydrogen vehicle and supply technologies to be commercially ready to serve the mass light duty vehicle markets in 20 years, development and demonstration needs to start now.

Questions

There are a number of questions raised by this analysis of externalities. For example, will using fossil-based hydrogen supplies on an interim basis "lock-in" hydrogen supply systems that might be incompatible with long term near-zero greenhouse gas emission routes? Would widespread use of distributed hydrogen production in small natural gas reformers interfere with future large-scale development of hydrogen from coal with CO_2 sequestration or from wind power? Will the use of fossil hydrogen without sequestration be on such a small scale globally that it won't make much difference in determining long term supply? What scale of development is needed to bring near-zero emission hydrogen end-use and production technologies to readiness for large-scale implementation? Can the development of hydrogen fuel cell vehicles be fueled with fossil-derived hydrogen at low cost and an acceptable level of primary energy use and emissions?

A Low Cost Hydrogen Transition: Matching Supply and Demand

To keep hydrogen costs relatively low during a transition, it is desirable to match hydrogen supply capacity fairly closely with the changing level of hydrogen demand. This avoids the expense of having underutilized hydrogen production plants and distribution systems. Certain types of hydrogen supply are well matched to various levels of demand. How does the level of demand influence the choice of hydrogen supply over time?

Observations

Hydrogen production systems appear to have "preferred" scales. For example, coal gasification plants with capture of CO_2 and nuclear hydrogen plants are projected to have better economics at very large capacity, on the order of 1000 Megawatts of hydrogen output, or several times the capacity of a large industrial hydrogen plant today. Such a plant could produce enough hydrogen to supply a fleet of more than 1 million efficient light duty vehicles. Biomass gasifiers also show scale economies favoring large size. However, biomass plant size can be limited by the high cost of transporting biomass long distances, a trade-off that leads to a lower optimal size for biomass plants. In contrast, water electrolysis and steam reforming of natural gas can be used to make hydrogen over a wide range of scales, from large central plants to small systems suitable for distributed hydrogen production at refueling stations serving a few tens to hundreds of vehicles.

Under some circumstances, distributed production of hydrogen at refueling sites might give a lower delivered hydrogen cost than centralized production with distribution. There is a trade-off between hydrogen production costs, which are generally lower in large scale production plants located near low cost feedstocks, and distribution costs, which are avoided for on-site production but can be large for centralized production. Distributed production options that rely on fossil fuels have the disadvantage that carbon cannot be readily captured for sequestration.

Centralized hydrogen production could have advantages for reducing emissions, especially with hydrocarbon feedstocks. With central hydrogen production and delivery, a wider variety of primary sources can be used than with on-site production. Management of CO_2 and other emissions from hydrogen production plants would be easier on a large scale, including capture of CO_2, which is not economically attractive on a small scale. Large scale production of zero carbon emission electricity could be used to produce electrolytic hydrogen at refueling sites.

Hypotheses

Matching a growing demand might require a succession of different hydrogen supplies, beginning with hydrogen delivered by truck or made via onsite electrolysis or steam reforming and moving toward dedicated central hydrogen plants with pipeline delivery in densely populated areas.

Matching supply and demand will probably require coordination between fuel suppliers and vehicle manufacturers. As noted in the previous section on externalities, to realize deep cuts in carbon emissions from the fuel sector in the longer term, it will be important to implement low carbon emission sources of hydrogen. Hydrogen from near-zero carbon emission supplies could be introduced to match a growing demand in several ways. In those locations where low-cost, zero-emission electricity is available, for example using off-peak hydropower, electrolytic hydrogen might be used in the early stages of a hydrogen infrastructure development. If low cost biomass feedstocks are available, hydrogen could be produced at intermediate scale with zero net carbon emissions. With large scale, central production, a variety of near-zero emission hydrogen production options could be used, including large renewable hydrogen systems, fossil hydrogen production with CO_2 sequestration, or nuclear hydrogen production. Hydrogen transportation fuel demand by itself probably won't be large enough or concentrated enough to support a large, dedicated hydrogen plant for several decades. If hydrogen is made on a large scale with zero emissions, primarily for industrial uses, some hydrogen could be distributed for use in vehicles.

Questions

How can supply change over time to match a growing demand at low cost without "stranded assets?" Is it possible to match zero emission hydrogen supplies to demand at a reasonable cost during a transition? When should a switch from low cost fossil sources to zero emission hydrogen sources begin? An important part of matching supply and demand is that hydrogen be delivered reliably when it is needed. As with today's energy carriers, this will require energy storage, back-up supplies, distribution options, and safeguards. What is needed to assure a reliable, secure supply of hydrogen? How does security depend on the primary supply used to make hydrogen and interdependence with other parts of the energy system?

Geographic and Regional Hydrogen Supply Issues

Many of the factors entering into the choices of hydrogen supply options are geographically specific. These include availability of primary and secondary resources for hydrogen production, the size and geographic density of hydrogen demand, the type of existing infrastructure, markets, and policies. How will geographic and regional issues affect the choice of hydrogen supply and the cost of building a hydrogen supply infrastructure?

Observations

Potential resources for hydrogen production are large and are more evenly distributed geographically than current oil resources. In many regions of the world, there are potentially large primary supplies that could supply hydrogen fuel via zero or near-zero emission pathways. These include renewable resources and fossil fuels with nearby sites for carbon sequestration.

Hydrogen demand is geographically specific. This affects how hydrogen is distributed and which primary energy supplies are used. Large demands in urban areas with high population densities will require a different distribution system than more sparse demands in rural areas. Finding the lowest cost hydrogen supply for a particular demand and location involves a trade-off between production and distribution costs.

Much of the projected growth in global demand for fuels and numbers of vehicles over the next 30 years is projected to take place in developing countries, where there are currently relatively few vehicles and little fuel infrastructure compared to industrialized countries.

Hypotheses

Depending on the region, different primary resources might be used to make hydrogen. In the long term, there will likely be a mix of primary resources for hydrogen supply, as with the electricity system today, and a mix of hydrogen distribution modes. The mix will probably change as demand grows, and the cost and availability of primary resources change over time.

Hydrogen will start as a regional fuel. In the United States, hydrogen vehicles will probably be used first in large urban areas with serious air quality problems, such as Los Angeles or New York City. A few areas in the world such as islands with significant renewable resources and expensive imported liquid fuels might be ideal first sites for hydrogen energy use.

Over the next few decades, large new transportation fuel supply infrastructures will be built to serve demand in developing countries. Hydrogen will not be entering a "green-field" in developing countries in just a few decades but will have to compete with or build upon the energy infrastructures now being put in place.

Questions

If hydrogen is widely used as an energy carrier, remaining unanswered questions related to geography will include the role of long distance transport of primary energy feedstocks for hydrogen production, and long distance transport of hydrogen fuel itself, in transportation energy supply. Will fuel supply become more regional in nature, more like the electricity system today? What role, if any, will hydrogen play in developing countries over the next few decades? For example, are air quality or oil supply security concerns in the cities of developing countries enough motivation to favor hydrogen vehicles over nearer term and lower cost low emission internal combustion engine technologies using conventional fuels? Will fuel infrastructures now being built in developing countries be compatible with future use of hydrogen? What are the opportunities to design new energy systems—for example, new natural gas pipelines—for future hydrogen compatibility?

Synergies of Hydrogen with the Existing Energy System

Existing infrastructure could provide a basis for developing a hydrogen supply system. Excess capacity in the current industrial hydrogen infrastructure might be used to provide hydrogen to early vehicle projects. Today's gasoline stations are well located to serve customers and might become sites of future hydrogen stations. Where they are available, secondary energy carriers such as natural gas and electricity can be used to make hydrogen at refueling sites, without the need for an extensive hydrogen distribution system. Future hydrogen pipelines might be built along existing rights of way for gas pipelines, electric transmission lines, major highways, or railroads. It has been suggested that the existing natural gas pipeline system might be used to carry hydrogen. As a near term strategy, hydrogen could be blended with natural gas at up to 15 to 20 percent hydrogen by volume to reduce air pollutant emissions without changing the gas pipeline distribution system. For higher fractions of hydrogen or pure hydrogen, the parts of the natural gas distribution system, such as seals and meters, and end-use systems, including engines or burners, would require changes. Existing primary resource extraction and delivery systems for natural gas and coal are well established to connect resources to conversion sites.

A future hydrogen supply system will be interdependent with the rest of the energy system, especially the electricity system. For example hydrogen might be produced from existing energy carriers such as electricity or natural gas. Co-production of hydrogen and electricity can improve the overall economics of the system. This is true for cogeneration systems in buildings where heat can be produced as well as for large-scale fossil hydrogen plants. It has been proposed that hydrogen could be used as a storage medium for intermittent electricity, such as solar or wind power.

The existing energy infrastructure could strongly influence how future hydrogen supply evolves in the near term. In the long term, some sites used for energy infrastructure today, such as rights of way, gasoline stations, and primary resource infrastructures, might remain in use for hydrogen systems. New development might also be required in the form of large scale renewable energy production facilities, fossil hydrogen production with CO_2 sequestration, and associated CO_2 disposal infrastructure. Moreover, fuel might be delivered to vehicles in new locations such as home or work. Development of hydrogen as an energy carrier might require cooperation between different industries in the energy system, fuel suppliers, and vehicle manufacturers, and also electric or gas utilities and chemical industries that use hydrogen.

Remaining questions involve the interaction of hydrogen with electricity and other parts of the energy system in a future system that relies on electricity, hydrogen, and possibly liquid fuels. Would use of hydrogen improve the reliability and resistance of the transportation fuel infrastructure to disruptions?

Looking to the Future for Hydrogen Supply

There are many possibilities for hydrogen production and delivery, including a number of promising zero or near-zero emission hydrogen supplies. The current lack of knowledge about future demand and markets for hydrogen vehicles complicates the problem of envisioning a transition and making projections about the timing for using various hydrogen supply options.

Nonetheless, several assertions can be made about future hydrogen supply. In the long term, to fully realize hydrogen's multiple benefits, it will be important to use hydrogen supply pathways with near-zero emissions of greenhouse gases and air pollutants, and to have the possibility of using diverse primary supplies. There are also many zero or near-zero emission hydrogen supply options, but these are currently more expensive than fossil hydrogen in most locations. Over the next few decades, while developing hydrogen end-use technologies such as fuel cells, it is important to simultaneously develop near-zero emission hydrogen production technologies that could be employed on a global scale. Promising long term options that could reach both low cost and zero or near-zero emissions include fossil hydrogen production with CO_2 sequestration, renewable hydrogen from biomass gasification or wind powered electrolysis, and hydrogen from off-peak power based on carbon-free electricity.

Natural gas might be an acceptable "compromise" transitional source of hydrogen over the next few decades in terms of low cost and low emissions. The greenhouse gas benefit of using hydrogen from natural gas in advanced hydrogen vehicles is small compared to fossil derived liquid fuels in improved internal combustion engine vehicles. However, there would be reduced emissions of air pollutants and reduced oil use (although greatly expanded use of natural gas in the United States might come from imports, bringing its own security issues). It might be possible to develop hydrogen vehicle technologies and bring them to technical readiness over the next few decades, fueled with hydrogen from natural gas, while achieving a reduction of societal impacts of energy, as compared to what might be achieved by advanced internal combustion engine vehicles. The impact of making hydrogen for the next decade or so on natural gas supply would be relatively small. Beyond a few decades, it would be necessary to change from natural gas without CO_2 sequestration to near-zero carbon hydrogen supplies. There is a debate about whether using natural gas to make hydrogen for a few decades would impede a later switch to lower carbon sources or whether it constitutes a bridge, allowing development of end-use systems using low cost hydrogen.

There are likely to be many solutions for hydrogen supply depending on the level of demand, resource availability, geographic factors, and progress in hydrogen production technologies. In the long term, there will be a mix of primary resources for hydrogen supply and hydrogen distribution modes. The mix will probably change as demand grows, and the cost and availability of primary resources change over time. Depending on the region, different primary resources might be used to make hydrogen. If externalities are highly valued, this will tend to favor near-zero emission hydrogen options. Hydrogen will develop first in regions where the case seems compelling on a policy–societal or economic basis.

The existing energy infrastructure could strongly influence how hydrogen supply evolves in the near term. In the long term, some sites used for energy infrastructure today might remain in use for hydrogen systems, but new development might also be required and new fuel delivery facilities built to allow refueling at home or at work.

A future hydrogen energy supply system will be interdependent with other parts of the energy system. It will be important to understand how hydrogen might fit, especially with the electricity and natural gas systems. Building a hydrogen supply will require coordination among many of the entities that have a stake in the transportation energy system, and possibly others, for example in the chemical industry. In terms of greenhouse gas emissions, we might have a few decades before we need to seriously decarbonize the fuel supply, but it is important to start now so that hydrogen and fuel cell technologies will be ready when and if we need them. Although hydrogen is not necessarily inevitable, it is one of the few options for deep reductions of greenhouse gases and pollutants from fuel use.

To go beyond these rather general statements about how hydrogen supply might evolve, much better knowledge of many highly uncertain, geographically specific, and time dependent factors would be needed. This includes knowledge that we do not now possess.

First, there is a need to better understand the potential demand for hydrogen. Here demand is broadly defined to include understanding future markets, economic competitiveness of hydrogen vehicles that might offer unique features to consumers with other advanced fuel and vehicle options, and geographic aspects of hydrogen demand. One of the greatest uncertainties here is that future hydrogen vehicles will face serious competition from other technologies such as extremely low-emitting, high-efficiency combustion vehicles. How cost effective will hydrogen vehicles be as a way of meeting environmental goals for transportation?

Second, a more certain knowledge of technical progress would be needed. Hydrogen technologies are advancing rapidly. Breakthroughs in hydrogen production, storage, or end-use technologies could change the way the market develops, and the way hydrogen is produced or distributed. Hydrogen markets and supply might evolve differently than envisioned here. Of course, hydrogen's competitors will be improving as well.

Finally, the extent to which externalities will drive the adoption of hydrogen is uncertain. Strong, consistent policies to simultaneously deal with air pollution, oil supply insecurity, and greenhouse gas emissions might accelerate the adoption of hydrogen technologies and zero or near-zero emission supplies. But other technologies that could also address these concerns would receive a "boost" as well. Although hydrogen offers the greatest potential reductions of externalities, the costs and benefits of hydrogen versus alternatives are not well quantified at present because of the uncertainties inherent in calculating the external costs of energy and the uncertain costs of future advanced vehicles.

Hydrogen offers huge potential benefits and a multiplicity of paths for realizing them. We have just begun a societal debate on the future role of hydrogen. The long term stakes are high enough that hydrogen is worth pursuing as an insurance policy, even if we don't yet know the answers to the many questions raised here. Finding the lowest cost and lowest emission solutions for hydrogen supply will be an ongoing work of engineering and design over the next decades, subject to many uncertainties and surprises. In the face of these uncertainties, it is not possible at present to make solid projections of future hydrogen demand or the speed at which a transition might proceed. To stimulate debate, it is possible to sketch evocative "existence proofs" of how a hydrogen system might look, how it might evolve, and its benefits and costs. These conceptual designs and plans might be guided by the general considerations sketched above.

References

Brown, L. C., G. E. Besenbruch, K. R. Schultz, A. C. Marshall, S. K. Showalter, P. S. Pickard, and J. F. Funk. (in press) "Nuclear production of hydrogen using thermochemical water-splitting cycles." Paper prepared for presentation at the American Nuclear Society Meeting Embedded Topical "International Congress on Advanced Nuclear Power Plants (ICAPP)," Hollywood, Florida, 9–13 June 2002. To be published in the *Proceedings of ICAPP*.

Burns, L., J. McCormick, and C. Borroni-Bird. 2002. Vehicle of change. *Scientific American* 287 (4): 64–73.

Energy Information Administration. 2003. *Energy Outlook with Projections to 2025*. Washington, DC: United States Department of Energy.

Gale, J., and Y. Kaya. 2003. *Greenhouse Gas Control Technologies*. New York, NY: Pergamon.

General Motors Corp., Argonne National Laboratory, BP, ExxonMobil, and Shell. 2001. *Well-to-Wheels Energy Use and Greenhouse Gas Emissions of Advanced Fuel/Vehicle Systems: North American Analysis, Executive Summary Report*. April, 2001. http://www.transportation.anl.gov/pdfs/TA/163.pdf.

International Energy Agency. 2003. *Greenhouse Gas R&D Programme Annual Report 2002*. Cheltenham, Gloucester, UK. http://www.ieagreen.org.uk.

Mintz, M. (Argonne National Laboratory) and M. Singh (Oak Ridge National Laboratories). 2003. Private communications.

Nakicenovic, N., A. Grubler, and A. McDonald, eds. 1998. *Global Energy Perspectives, Institute for Applied Systems Analysis and the World Energy Council*. Cambridge, UK: Cambridge University Press.

Ogden, J. M., R. H. Williams and E. D. Larson. 2004. A Societal Lifecycle Cost Comparison of Alternative Fueled Vehicles. *Energy Policy* 32: 7–27.

Thomas, C. E., I. F. Kuhn, B. D. James, F. D. Lomax, and G. N. Baum. 1998a. Affordable Hydrogen Supply Pathways for Fuel Cell Vehicles. *International Journal of Hydrogen Energy* 23(6): 507–516.

Thomas, C. E., B. D. James, F. D. Lomax, and I. F. Kuhn. 1998b. *Draft Final Report, Integrated Analysis of Hydrogen Passenger Vehicle Transportation Pathways*. Report to the National Renewable Energy Laboratory, Under Subcontract No. AXE-6-16685-01. Golden, CO: U.S. Department of Energy, March 1998.

U.S. Department of Energy. 2003a. "Carbon Sequestration: Technology Roadmap and Program Plan." Washington, DC: U.S. Department of Energy, Office of Fossil Energy, National Energy Technology Laboratory. March 12, 2003.

U.S. Department of Energy. 2003b. *Hydrogen Fuel Cells, and Infrastructure Technologies Program*. Washington, DC: U.S. Department of Energy, Office of Energy Efficiency and Renewable Energy. http://www.eere.energy.gov/hydrogenandfuelcells/hydrogen/production.html.

Wang, M. 2002. "Fuel Choices for Fuel Cell Vehicles: Well-to-Wheels Energy and Emission Impacts." *Journal of Power Sources* 112: 3017–321.

Wang, M. Q. 1999. "GREET 1.5—Transportation Fuel-Cycle Model 1 Methodology, Development, Use and Results, Report No. ANL/ESD-40." Center for Transportation

Research, Argonne National Laboratory, prepared for the Office of Transportation Technologies, U.S. Department of Energy. Washington, DC: U.S. Department of Energy, December 1999.

Weiss, M., J. Heywood, E. Drake, A. Schafer, and F. AuYeung. 2000. "On the Road in 2020." MIT Energy Laboratory Report # MIT EL 00-003. Cambridge, MA: Massachusetts Institute of Technology, October 2000.

Wigley, T. M. L., R. Richels, and J. A. Edmonds. 1996. Economic and Environmental Choices in the Stabilization of Atmospheric CO_2 Concentrations. *Nature* 379(6562): 240–243.

Williams, R. H. 2002. "Decarbonized Fossil Energy Carriers and Their Energy Technological Competitors." Prepared for the IPCC Workshop on Carbon Capture and Storage, Regina, Saskatchewan, Canada, November 18–21, 2002.

CHAPTER 6

Clean Hydrogen from Coal with CO_2 Capture and Sequestration

Richard D. Doctor and John C. Molburg

Most near term hydrogen will likely be made from natural gas. A domestically secure alternative is coal. Every component of coal-to-hydrogen production processes is not only demonstrated on a commercial scale but is in commercial operation outside the United States, for example, to produce ammonia in China. The economics of large scale coal-to-hydrogen looks favorable if the 2003 prices for natural gas remain the norm. The largest challenges will be the transport of carbon dioxide and the availability of carbon dioxide sequestration reservoirs. Here the regulatory landscape is still not fully formed.

This chapter examines the opportunity for coal-derived hydrogen production technologies, with special emphasis on environmental considerations. It describes a state-of-the-art coal-to-hydrogen demonstration planned by the U.S. Department of Energy (DOE), a project now underway in North Dakota, and it explores the potential role of coal in a hydrogen energy economy of the future.

Steam Reforming of Natural Gas and Petroleum for Hydrogen Production

What will be the demand for hydrogen in a hydrogen economy, and where will this hydrogen come from? One demand scenario for the U.S. hydrogen economy posits an annual hydrogen production rate of 20-90 million tons (National Academy Press, 2003). A posited 2030 level of 40 million tons annual hydrogen production would be sufficient to power 150 million light duty vehicles averaging the gasoline-equivalent of 60 miles per gallon (mpg).

Several resources could be used to produce the hydrogen. Leading contenders include fossil, renewable, and nuclear energy.

At present in the United States, the resource chain for producing hydrogen starts with petroleum products, such as naphtha, or with natural gas. These feedstocks are converted to hydrogen and carbon dioxide by means of a process called catalytic reforming. Variations on the basic reforming process account for 96 percent of all hydrogen produced today. Worldwide, 63 percent of the hydrogen produced is used for manufacturing ammonia fertilizers, with the remainder principally used in refineries to remove sulfur and nitrogen pollutants from gasoline, diesel, and jet fuel (Hairston, 2001). Current U.S. and global hydrogen markets are shown in Fig. 6-1 compared to the posited 2030 U.S. hydrogen demand of 40 million metric tons.

For a methane feed, the overall summary reaction for the two-step process is:

$$CH_4 + 2H_2O = 4H_2 + CO_2 \quad \Delta H \text{ @ } 298.15°K = 113 \text{ kJ/kg-mole}$$

This reaction takes place over two or more beds of catalyst and is endothermic; that is, it requires an input of heat. Usually, the required heat energy comes from partial or complete combustion of some of the natural gas. The unconverted gas remaining after hydrogen purification is also used as a furnace fuel. Approximately 40 percent of the natural gas by weight is required to supply the heat to drive the reforming reaction.

Annual natural gas use in the United States during 2001 was nearly 616 billion normal cubic meters (23 trillion standard cubic feet). Using

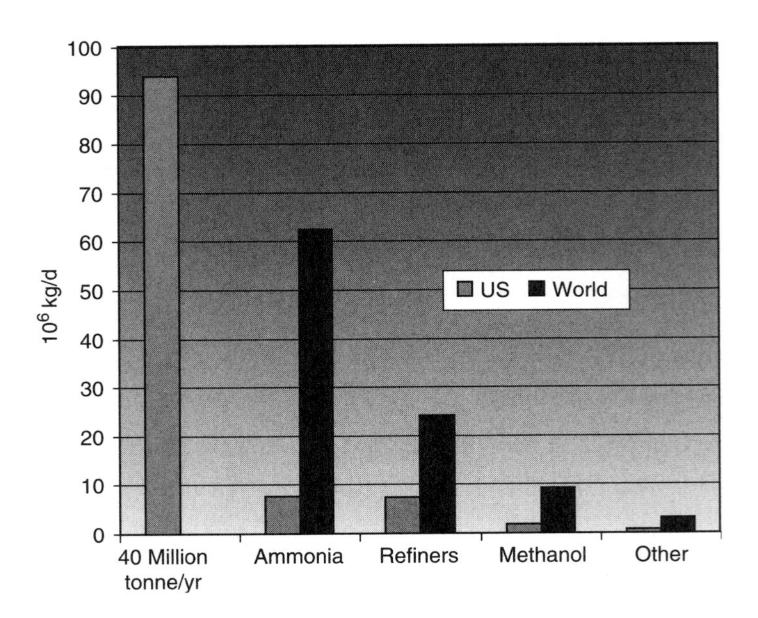

FIGURE 6-1. Target hydrogen of 40 million metric tons per year compared against U.S. and world production.

natural gas as a feedstock to produce the required 40 million metric tons (t) of hydrogen per year would require a 27 percent increase in the natural gas supply. This would be in addition to increases in natural gas use to produce electric power for home and industrial applications. Such an approach would require a large investment, not just in a hydrogen infrastructure but also in an expanded natural gas infrastructure. Natural gas exploration, and the production or importation of liquefied natural gas (LNG), would have to expand to meet the increased market in order to maintain economics competitive with those of today's technology. The U.S. LNG infrastructure can currently manage 21 billion normal cubic meters (0.8 trillion standard cubic feet) annually. If this use were massively expanded, the results would include increased dependence on foreign sources of primary energy. The U.S. chemical industry currently uses 12 percent of the nation's natural gas production as a feedstock. Rising natural gas costs are making the domestic production of basic petrochemicals for plastics uneconomical (Hairston, 2003). It is clear that demands for hydrogen feedstock would increase the pressure on the price of natural gas, exacerbating a situation that already is being termed a "crisis" by an embattled chemical industry.

The Opportunity for Coal-Derived Hydrogen

The challenge to the continued use of natural gas creates an opportunity for coal-derived hydrogen. On a worldwide basis, approximately 18 percent of the world's hydrogen already is derived from coal (Kirk-Othmer, 1995). At the current rate of consumption, the United States has more than a 300-year supply of coal, with the primary reserves being the high-sulfur, high-heating value coal available in the Midwest and Appalachia. Of particular importance is the low-sulfur, low-heating value coal from the Powder River Basin in Wyoming that now accounts for about 30 percent of U.S. coal consumption.

Producing hydrogen from coal is more capital intensive than producing it from natural gas. However, this approach deserves serious consideration because the costs of coal are lower and are likely to remain so even with significant increases in demand. Coal is domestically abundant, is secure in terms of delivery, and has never shown the market volatility of natural gas.

The Great Plains Gasification Project—located in Beulah, ND, and now owned by Basin Electric, headquartered in Bismarck, ND—employs 14 Lurgi coal gasifiers in a $2.2 billion (1984 prices) project that has been in operation for nearly twenty years (Fig. 6-2). More advanced entrained-flow gasifiers dominate today's market, so this facility is not state of the art.

The plant, which began operating in 1984, produces more than 1.5 billion normal cubic meters (54 billion standard cubic feet) of synthetic natural gas annually. Coal consumption exceeds 15,000 t each day. Synthetic natural gas leaves the plant through a 610 mm (2-ft)-diameter pipeline, traveling 34 miles (54 km) south. There it joins the Northern Border Pipeline, which transports the gas to four pipeline companies. In addition to natural gas,

FIGURE 6-2. Dakota Gasification Company and the Great Plains Synfuels Plant. *Source:* Great Plains Gasification Project, 1987.

Great Plains also produces fertilizers, solvents, phenol, carbon dioxide, and other chemicals. The CO_2 is now used in an international venture for enhanced oil recovery, and approximately 4400 t/day of recovered CO_2 is transported 215 miles (344 km) north to PanCanadian's Weyburn oil field. If the intermediate product gas were used strictly for hydrogen production, the yield would be 541 t/day, which is the energy equivalent of 13,000 barrels/day (bbl/day) of gasoline.

The only coal-to-hydrogen plant in the United States is the Tennessee Eastman Chemical Facility. However, the U.S. oils refining industry has been gaining considerable experience in gasification technology that could be applied to coal-to-hydrogen plants. Gasification is now used to convert excess petroleum coke to electricity for use in those same refineries. The same basic process could take the synthesis gas from the gasification and use it to produce hydrogen instead of electricity.

Coal is a potentially more economical source of hydrogen than natural gas in developing economies such as China, which do not have the extensive natural gas distribution infrastructures as in the United States. The market for early coal-to-hydrogen systems in China has been dominated by ChevronTexaco and Royal Dutch Shell. Typical of the sort of financial arrangements emerging today are Shell's joint venture with Sinopec to construct a $140 million, 2000 t/day coal gasification plant at the Dong Ting fertilizer plant in Hunan Province. The plant will convert coal into hydrogen, replacing the current feedstock, naphtha. There is no CO_2 recovery proposed for this plant (Royal Dutch Shell, 2003).

The U.S. Department of Energy's Office of Fossil Energy is soliciting industrial interest in the FutureGen plant, a near-zero emissions coal-fed plant that will produce power and hydrogen (White House, 2003; U.S. Department of Energy, 2003). Unlike current coal-to-hydrogen technologies, FutureGen plants would employ integrated CO_2 management. The anticipated $1.0 billion budget will support the design, construction, and operation of a 275 megawatt (MW) prototype plant to serve as a large-scale engineering laboratory for testing new clean power, carbon capture, and coal-to-hydrogen technologies. It will be the cleanest fossil fuel-fired power plant in the world, with virtually every aspect of the prototype plant employing cutting-edge technology. A three-part effort is envisioned, comprising core research and development, infrastructure development, and plant integration. With respect to CO_2 mitigation, captured CO_2 will be permanently sequestered in a geological formation. Candidate reservoirs include depleted oil and gas reservoirs, unmineable coal seams, and deep saline aquifers all common in the United States.

Full Energy Cycle Analysis of Coal-to-Hydrogen

Doctor, *et al.* (2001) summarizes a full energy cycle analysis (Fig. 6-3), which combines the process design and economics of CO_2-capture technologies used with an integrated gasification combined-cycle (IGCC) power system. Within the plant, commercially available CO_2-capture technology provides the performance and economic baseline against which to compare innovative technologies.

The full energy cycle approach recognizes that there are process impacts beyond the plant boundary. An important consideration in this energy cycle analysis is the derating of the power plant resulting from CO_2 capture and sequestration. An equivalent CO_2 charge must be assessed for

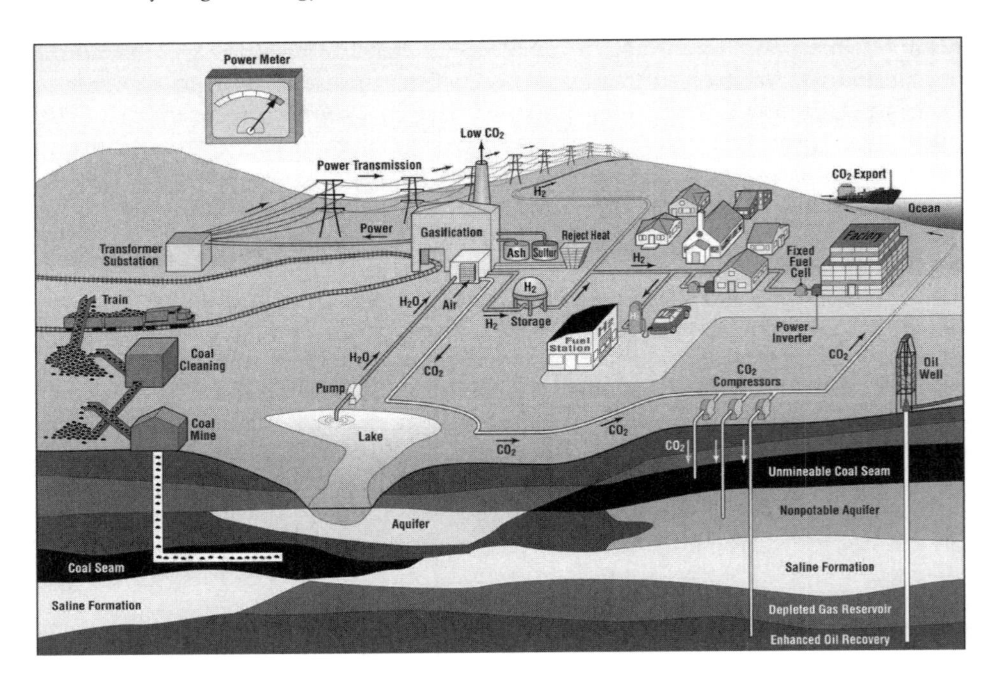

FIGURE 6-3. A "full energy cycle analysis" that combines the process design and economics of CO_2-capture technologies with integrated gasification combined-cycle (IGCC) power systems and hydrogen production.

makeup electricity at about 1 kg CO_2 per kilowatt hour (kWh). The full energy cycle includes:

- Mining a deep seam of Illinois #6 coal
- Coal transport by rail for 160 km
- Coal preparation, sizing and drying
- The IGCC plant

The basis for the plant design is a 450-MW (gross) IGCC system with an O_2-blown Shell entrained-flow gasifier, Illinois #6 bituminous coal feed, and low-pressure glycol sulfur removal. This is followed by Claus/SCOT treatment to produce a saleable sulfur product. Then 5500 t/day of CO_2 is delivered at supercritical pressures. Hydrogen undergoes final cleanup using pressure swing adsorption to prepare it for delivery. The 2700 t/day of high-sulfur coal feed is converted to 322 t/day of hydrogen, or approximately 7700 bbl/day gasoline equivalent. The major process steps required are shown in schematic form in Fig. 6-4.

While commercial conversion of natural gas to hydrogen without CO_2 capture exhibits a lower heating value (LHV) efficiency of 77 percent, converting coal to hydrogen with entrained-flow gasifiers and CO_2 capture

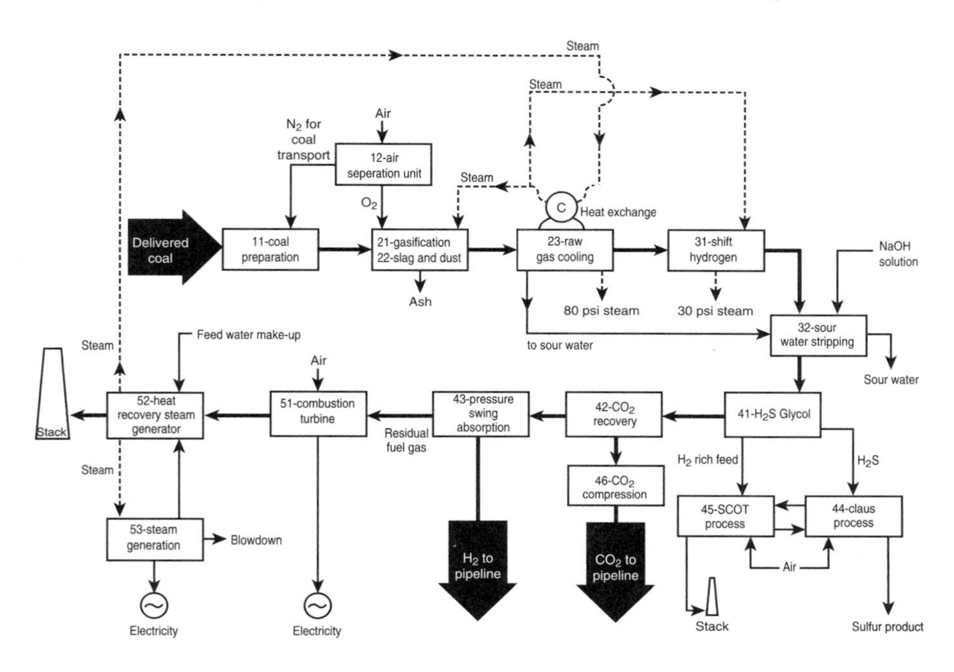

FIGURE 6-4. Current design for high-sulfur coal-to-hydrogen and electricity with CO_2 capture.

using either Shell or ChevronTexaco technology should exhibit LHV efficiencies of roughly 55 percent. Lower heating value means that the heat of the water vapor is not recovered, while higher heating value means this heat is recovered, a distinction important for efficiency claims. HHV heat recovery has never proven practical for these facilities.

If the price of hydrogen is set by the market price of natural gas and CO_2 sequestration is required, then coal-to-hydrogen becomes competitive when the long-term price for natural gas reaches \$4.00 to \$4.50 per million Btu. This means the plant gate costs of hydrogen are \$1.30 to \$1.40/kg (1 kg hydrogen has about the same energy as one gallon of gasoline). It is important to emphasize that all the process steps studied have already been demonstrated at a full commercial scale.

CO_2 Transportation

Once the coal is converted to hydrogen, captured CO_2, and electricity, the CO_2 must be transported to the site of sequestration. CO_2 can be transported as a compressed gas, a liquid, a solid, or in a high-pressure supercritical state. If a pipeline right-of-way is available, economic considerations show a clear advantage for CO_2 transport in the supercritical state; pipelines delivering nearly 50,000 t/day of supercritical CO_2 have operated safely since the 1980s. The technical specifications for these pipelines show that they are comparable to those employed for carrying petroleum and natural gas, although they operate at somewhat higher pressures. Some pipelines are over 600 km in length and 550 mm in diameter, and operate at pressures as high as 150 bar.

Issues such as achieving the required purity of the CO_2, transfer of custody, pipeline maintenance, and safety are well understood, and thus land-based transport of CO_2 must be considered as a mature technology. An individual CO_2 pipeline that is safe, reliable, and economical can be constructed with assurance (Fig. 6-5).

At the same time, the deployment of an infrastructure for carrying CO_2 in sufficient quantities to meet anticipated demands for climate mitigation shows that the current experience must be extrapolated cautiously. Right-of-way clearance may become more difficult to obtain. This will particularly be true in highly populated regions with high CO_2 releases. Safety issues will become quite complex within the United States where CO_2 transport is treated as a hazardous material, but not declared hazardous. All U.S. review requirements for high risk hazardous pipelines are applied if the pipeline is greater than 457 mm (18 in.) or passes through a

FIGURE 6-5. Cortez CO_2 pipeline with custody transfer meter; this is a 640 km (400 mile)-long pipeline delivering CO_2 to the Texas Permian Basin for enhanced oil recovery. *Source:* Kinder-Morgan Company, LP.

populated area, meaning a population density greater than 1000/square mile (386/square km). Current experience has thus far been in areas with low population densities.

The most serious danger that a pipeline faces is a rupture caused by uninformed digging. More pipelines mean more accidents. The tendency for CO_2 to remain at ground level or move into low areas creates novel safety risks that need to be evaluated for each CO_2 transport system at each site. Ultimately, insurance and liability issues may prove to be a constraint on CO_2 transport.

The most economical transport systems favor CO_2 capture from centralized power plants, where CO_2 volumes will be large. The smaller power plants now commonly being built to produce electricity from natural gas make CO_2 capture and transport much more costly, suggesting the likelihood of stranded CO_2 at such sites.

There will be pressure for international transport of carbon dioxide because massive, secure reservoirs may well not be available in many countries. This will mean the development of economical and secure oceanic shipment of CO_2. Again, this will build on an existing base of experience, but one that is at present limited to the operation of three ships throughout the world.

Hence, a regulatory framework to balance these issues and provide a framework for the low greenhouse gas emissions power industry of the future to make investment decisions will need to emerge. Future power plant owners may find the CO_2 transport component one of the leading issues in their decision making.

CO_2 Sequestration

Sequestration options are to place CO_2 underground into various geological formations, convert it to solid mineral forms, or create agricultural and forest reservoirs that use photosynthesis in trees, grasses, and crops to convert atmospheric carbon dioxide. Our current understanding of CO_2 sequestration has been summarized in a critical review by the National Energy Technology Laboratory (NETL) (White *et al.*, 2003). At present, the domestic capacity of some of the most promising reservoirs (forests and soils >100 gigatons capacity; geological 300-3200 gigatons capacity; oceans 1400-20 million gigatons capacity) still must be refined, but clearly there are significant opportunities.

On a geological time scale, the important biologically mediated, permanent sequestration of CO_2 does not occur on land. It is in the oceans where limestone is formed that we would naturally seek the massive reservoirs of sequestered CO_2. Because the oceans will eventually be the natural sink for CO_2, concepts for directly sequestering CO_2 in the oceans have been proposed. But ocean sequestration and ocean fertilization have proven to be extremely controversial and politically unpopular. Proposed demonstration experiments to show the feasibility of CO_2 ocean sequestration have thus far been rebuffed in Hawaii, Norway, and Japan (Gewin, 2002). Any use of the oceans as a direct CO_2 sink will require considerable international negotiation. This sequestration opportunity is clearly not ready for near-term deployment, nor may it ever be ready.

Efforts to turn CO_2 into carbonate rock on land—mimicking the manner in which the ocean biosphere has managed carbon for 3.6 billion years—would also prove costly. Proponents of taking a 100,000-year natural cycle and compressing it into one hour are projecting high costs as they develop serious engineering estimates, but these investigations are continuing and may lead to a breakthrough (Lackner, 2003; Lyons, 2003).

On the other hand, geological sequestration is a currently available technology. Exploiting this opportunity will require developing two new international industries: one to recover and transport CO_2 via pipelines and a second to manage the new reservoirs. The reservoir management industry would hold the liability for sequestering or converting CO_2 so that it can be contained either on land or in the sea in geologically secure forms, on a timescale of tens to hundreds of thousands of years.

In addition to carbon sequestration, technologies that would provide economic benefits include those that enhance oil recovery, produce coalbed methane, and maintain pressures in depleted gas reservoirs to avoid surface subsidence. Currently, companies in the United States sell one billion standard cubic feet of CO_2 each day, or approximately the CO_2 output from one conventional coal-fired electric power plant with a power capacity of 2300 MW. This CO_2 is used economically and with little or no environmental impact for approximately 70 enhanced oil recovery projects and for other industrial applications. Pipeline specifications for CO_2 quality, pipeline safety issues, and custody of the CO_2 have a base of industrial experience that goes back to the 1970s. Today, there are operating CO_2 pipelines of up to 760 mm (30 inches) in diameter and 640 km (400 miles) in length (Fig. 6-6).

Carbon Taxes and CO_2 Sequestration

Norway, with a tax as high as \$54.00/t of CO_2, is currently gaining attention for its massive sequestration project, the sole purpose of which is to avoid atmospheric emissions. At the Sleipner Field, Statoil of Norway separates CO_2 associated with natural gas production from even deeper geological strata at the same site. Statoil annually injects one million metric tons of CO_2 into a capped geological formation 1000 meters below a shallow sea inlet. To provide a basis for comparison, this is approximately the CO_2 output from a 120-MW electric coal-fired power plant.

The volume of economically beneficial or neutral uses of CO_2 is a fraction of the carbon produced by the current electric power industry in the United States, let alone the transportation sector. Total U.S. emissions of CO_2 are around 6.2 billion metric tons annually.

FIGURE 6-6. Cortez CO_2 pipeline under construction. *Source:* Kinder-Morgan Company, LP.

In contrast, the potential for CO_2 sequestration is enormous. The storage capacity of CO_2 in coal beds ranges from 225 to 964 billion metric tons. Brine fields in the United States have a capacity as high as 550 billion metric tons, while worldwide capacity might be 350 to 11,000 billion metric tons. Hence, the geologic potential for large-scale permanent sequestration opportunities is significant.

Conclusions

The coal-to-hydrogen opportunity is real, and coal-derived hydrogen prices probably will not be subject to the market fluctuations that characterize natural gas prices. Coal-to-hydrogen production is already competitive elsewhere in the world, although economic considerations will dictate higher capital investments for coal-based systems than for steam methane reforming. Capture of CO_2 and domestic transport of CO_2 already have an industrial base of experience that goes back to the 1970s. With respect to CO_2 sequestration, estimates of reservoir use and capacities are large but not well established. The geologic potential for large scale sequestration exists, but public acceptance and therefore sequestration costs are highly uncertain.

References

Doctor, R. D., J. C. Molburg, N. F. Brockmeier, and G. Stiegel. 2001. "Designing for Hydrogen, Electricity, and CO_2 Recovery from a Shell Gasification Based System." Newcastle, New South Wales, Australia: 18th Annual International Pittsburgh Coal Conference, December 4–7, 2001.

Energy and Transportation Challenges for the Chemical Sciences in the 21st Century, Organizing Committee for the Workshop on Energy and Transportation, Committee on Challenges for the Chemical Sciences in the 21st Century, National Research Council, National Academy Press (2003): 68.

Gewin, V. 2002. Ocean Carbon Study to Quit Hawaii. *Nature* 417 (June 27): 888.

Hairston, D. 2001. The Utility of Hydrogen, *Chemical & Engineering News,* 108(9):29.

Hairston, D. 2003. Natural Gas Crisis Looms. *Chemical & Engineering News* 81(23): 7.

Lackner, K. 2003. A Guide to CO_2 Sequestration. *Science* 300 (June 13): 1677.

Lyons. 2003. "Hydrogen—Manufacture." *Kirk-Othmer Encyclopedia of Chemical Technology,* 4th ed., Vol. 13. New York: Wiley & Sons, 1995: 838–894.

Royal Dutch Shell Oil. "Shell in China." http://www.shell.com.cn/english/environment/sustainable_1.html. (accessed August 28, 2003).

U.S. Department of Energy, Office of Fossil Energy, National Energy Technology Laboratory. "FutureGen." http://www.netl.doe.gov/coalpower/sequestration/futureGen/main.html. (accessed August 29, 2003).

White, C., B. Strazier, E. Granite, J. Hofman, and H. Pennline. 2003. "Separation and Capture of CO_2 from Large Stationary Sources and Sequestration in Geological Formations—Coal Beds and Deep Saline Aquifers." *Journal of the Air & Waste Management Association* 53(6) (June): 645.

White House. 2003. White House Press Release. February 27, 2003. http://www.whitehouse.gov/news/releases/2003/02/20030227-11.html. (accessed August 29, 2003).

CHAPTER 7

Doing Good by Doing Well: Entrepreneurship in the Hydrogen Transition

David L. Bodde

There is nothing more difficult to take in hand, more perilous to conduct, or more uncertain in its success, than to take the lead in the introduction of a new order of things.

Nicolo Machiavelli

The United States is now embarked upon a policy that would transform the national transportation fuel infrastructure to deliver energy-based goods and services with hydrogen as the energy carrier. Much analytic effort has been expended on the "what" questions—what environmental and security benefits a mature hydrogen economy would convey. Less thought has been devoted to the "how" questions—especially how the national fuel infrastructure, which performs reasonably well from a consumer perspective and still holds untapped potential for improved efficiency, can be induced by policy to make an early transition to a hydrogen-based economy. This chapter examines one course that transition policies might take—the nurturing of entrepreneurial activity and private investment in hydrogen-related businesses.

Current Policy Focus: Transportation

The benefits sought from a hydrogen-based energy economy are largely public: (1) reduced carbon emissions into the atmosphere, and hence a lower likelihood of disruptive climate change, and (2) improved energy security through a more diverse fuel supply, especially one that relies less on oil, much of which must be imported from politically volatile regions. These announced goals intersect in the transportation sector.

With regard to carbon emissions, transportation activities in the United States released 515 million tonnes (t) of carbon into the atmosphere in 2000, of which 503 million tonnes came from petroleum. Coal burning released another 522 million tonnes in the same year in the production of electricity. Taken together, petroleum use in transportation and coal burning for electricity account for the bulk of carbon emissions from the U.S. economy (Energy Information Administration, 2001a).

With regard to petroleum, about two-thirds of the oil consumed in the United States goes to the transportation sector. And within the transportation sector, essentially all the energy consumed (97 percent in 2002) comes from petroleum. Of this, over half must be imported, much coming from the most politically unstable regions of the globe.

Thus, the public case for hydrogen use is strongest in transportation, especially road transportation, and as a consequence much of the attention of the U.S. Department of Energy (DOE), the federal agency charged with implementing the hydrogen transition, focuses there. Transportation, however, presents an unusually difficult challenge for hydrogen-using technologies.

The "How" Problem

For the public benefits of hydrogen to be achieved, technologies for its production and use must be diffused throughout the economy. Yet, the government is neither the customer nor the producer. Instead, it must persuade an economy that allocates resources largely through markets to adopt the technology. In a free society, this means that consumers making informed choices must find hydrogen-based products and services attractive relative to the alternatives.

The established fuel infrastructure offers a challenging target indeed. The energy system for road transportation—fuels plus vehicles—has grown into an efficient, mature, and interlocking infrastructure. It is efficient in that affordable energy services are available to most people at most times. The retail price of motor gasoline, for example, has remained remarkably constant from the mid-1980s until 2002, averaging around $1.50 per gallon as measured in 1996 dollars. Resource cost is now less than half the retail price of gasoline. For example, the average retail price of gasoline was $1.48 in the United States in 2000. The cost of crude oil accounted for only 46 percent of this. Federal and state taxes accounted for 28 percent of the price, in contrast with the costs and profits for refining, distribution, and marketing at 26 percent (Energy Information Administration, 2001b). Though this performance might seem less satisfactory when the externalities of climate change and national security are included in the calculus, from the perspective of most consumers, the system "ain't broke."

The road transportation system is mature in that the rate of radical innovation (as distinct from incremental improvement) has slowed. And, it is interlocking in that technological change in one part of the system (say, engines in cars) pervades other parts of the system (including fuels, car design, roads, and so forth) unless the change is either incremental in nature or can otherwise be confined to one part of the system. The emerging generation of hybrid electric vehicles, for example, makes use of the extant fuels infrastructure without any requirement for conforming changes, and so the consequences of this innovation are confined to one part of the combined fuel and vehicle system. In contrast, system-wide innovations like a hydrogen-based fuel cell vehicle require the entire fuel–vehicle infrastructure to change, and might also require some modification to the way consumers use these vehicles. In general, such innovations find it difficult to penetrate interlocking, mainstream markets because their benefits must be compelling enough to overthrow the entire system—a condition rarely demonstrable in advance.

Compounding these difficulties, the current fuel and vehicle infrastructure shows an underused capacity for improvement. For example, a National Research Council committee reported that fuel economy improvements ranging from 12 to 48 percent could be achieved from technologies now within reach. These would not compromise key vehicle characteristics like size or acceleration, and all such investments would pay for themselves within the assumed 14-year life of the vehicle. In the case of large vehicles, the payback time could be as short as 3 years (National Research Council, 2002).

Taken together, the combination of an interlocking, efficient fuel and vehicle infrastructure, apparent consumer satisfaction with its performance, and meaningful potential for improvement using currently available technologies presents an enormous challenge for hydrogen in penetrating the transportation market. To understand how entrepreneurial activity might accelerate this, we must first examine what customers are actually seeking.

A Service Concept for Technology

An old marketing cliché holds that customers do not buy quarter-inch drills, they buy the expectation of quarter-inch holes. Similarly, fuels and vehicles have little value in themselves, but enormous utility as providers of mobility services. The most successful entrepreneurs think of their offering as a bundle of services and ask which among them will be highly valued by customers. These valued services form an n-dimensional space, which, for road transportation, might include performance vectors like:

- Time saving: Will the vehicle travel far enough that the driver does not waste time perpetually refueling?
- Economy: What does the vehicle cost to own and operate?
- Safety: How well does the vehicle protect the customer and the customer's family?

- Comfort: Can drivers and passengers arrive at their destination unstressed and unhassled?
- Image: What does driving this particular vehicle say about its owner?

Thus, the competitive battle is never about technologies; rather, it concerns the package of services that competing technologies offer. The general patterns of innovation observed throughout the economy reflect the outcomes of these competitions.

An Innovation Perspective

The term "innovation" does not describe a unified or even coherent process for introducing improved goods and services into a market economy. Nevertheless, studies of innovation in a variety of industries reveal general patterns, and these can illuminate the possibilities for a transition to hydrogen. A synthesis of these studies provides an economic demography, showing the central tendencies of the innovation process and how these might apply to a hydrogen transition.

As a first approximation, we can sort the universe of goods produced into two categories. The first is assembled products, manufactured goods that are built from discrete parts. Automobiles, fountain pens, and personal computers are among these, and their production is measured in units. In contrast, nonassembled products often become part of assembled products—the paint on the car, the glass on the windshield, the gasoline in the tank, and so forth. Nonassembled products can often be expressed as a chemical formula and are usually measured in terms of weight or volume rather than units.

Innovation comes to both assembled and nonassembled products in one of two ways. The innovation can be radical, offering a major departure from current practice and a discontinuous leap in some aspect of performance. Or it can be incremental, a series of small improvements, often undocumented, that accumulate to large gains. Incremental innovation underlies the continuous improvement made famous by Japanese manufacturers or the learning curve well known around the world.

The relationships among these aggregations of product and innovation types can be seen in the innovation matrix shown in Fig. 7-1. The horizontal axis represents the two classes of innovation, and the vertical axis the two classes of product. Each cell in the matrix then sets out an innovation space with unique characteristics.

Hydrogen is an archetypal nonassembled product, and the devices that convert it to services desired by consumers are archetypal assembled products. Thus, hydrogen use in transportation requires innovations at a roughly matching pace on both the top row and the bottom row of the innovation matrix, shown for the hydrogen transition in Fig. 7-2. Stimulating these innovations for the public purpose presents the DOE with a significant

	Radical innovation	Incremental innovation
Assembled products	▪ Mostly product innovations ▪ High rates of product change ▪ Can change: ○ Business model for producers ○ Use patterns for customers ▪ New startup companies	▪ Mostly process innovations ▪ Constrained product change ▪ Reinforces extant business models ▪ Established companies
Non-assembled products	▪ Occurs rarely ▪ Mostly process innovation ▪ Innovations *changing* business model come from outsiders ▪ Innovations *sustaining* business model come from insiders	▪ Occurs frequently ▪ Mostly process innovation ▪ Continuous improvement from insiders

FIGURE 7-1. Innovation matrix: central tendencies for radical or incremental innovation.

	Radical innovation	Incremental innovation
Assembled products	▪ Fuel cell vehicles ▪ Hydrogen storage systems ▪ Refueling devices ▪ Hydrogen detection systems	▪ Hybrid-electric vehicles* ▪ Natural gas vehicles*
Non-assembled products	▪ Distributed production at refueling sites: ○ New technology, new business model ▪ On-vehicle production: ○ New technology, old business model	▪ Steam reforming of methane

Innovation matrix for the hydrogen transition

*Competitors for hydrogen vehicles

FIGURE 7-2. Innovation matrix for the hydrogen transition.

challenge because the characteristics of the innovation process vary markedly among the cells in the matrix.

Radical Innovation in Assembled Products

In the top left quadrant of Fig. 7-1, radical innovations occur frequently in assembled products during the formative stages of their lifecycle. Here we find the most striking change in the service package available to consumers. In most cases, these radical innovations are brought into the market by entrepreneurs who recognize the value they can create and build a company to capture that value. Large companies, however, often serve as fast followers and bring the innovation to mass markets through greater manufacturing and marketing capabilities.

Consider the personal computer (PC), for example. The initial market was developed by Apple Computer, which first created a functional tool from what had been a hobbyist's toy. But Apple was slow in moving this innovation into the mainstream of business users and left it for IBM to introduce the PC into mass markets in 1981. That innovation began a transition that changed forever the nature of information services enjoyed by customers, whose work lives and personal possibilities were irrevocably altered. Over the next 20 years, the basis of competition for all involved in the computer industry—computer makers, software companies, and microprocessor makers—shifted from advanced technology to lowest cost, and eventually to services, frequently the end point of the commoditization process.

In the case of hydrogen, illustrated in Fig. 7-2, the radical innovations include fuel cells, refueling systems, hydrogen detection devices, and so forth. All of these component innovations must be integrated into the fuel cell vehicle design, and a significant underperformance in any one component—onboard storage, for example—could render the entire vehicle system uncompetitive in the marketplace. Much entrepreneurial activity has already occurred here, but its future rate and direction remain unclear.

Incremental Innovation in Assembled Products

Incremental innovations tend to dominate once the product has matured and the design parameters become stable. The most meaningful innovation occurs in the manufacturing process rather than the product itself. These process innovations are seen by the consumer largely as lower cost, and with time become quite significant. The product itself becomes more commodity-like with maturity, and product differentiation comes from the nontechnological vectors of performance—the "Swatch," for example, the fashionable timepiece that the Swiss watchmakers used to recapture some of the electronic watch market from the Japanese. Established companies, which perhaps grew from entrepreneurial startups, dominate here because of their superior marketing channels and manufacturing.

Two interim technologies—the hybrid electric vehicle and the natural gas vehicle—are shown in the top right quadrant of Fig. 7-2 because of their dual-edge relationship to the hydrogen transition. On the one hand, these are likely to provide formidable competition for the hydrogen fuel cell vehicle. Their costs will drop, and the services they offer will benefit from numerous incremental improvements as manufacturers gain experience and early adopters provide market feedback. Such improvements will also reinforce the chief performance differential that buyers of these interim vehicles might enjoy relative to conventional autos: lower driving cost and the satisfaction of driving a machine with superior environmental performance. On the other hand, hybrid electric and natural gas vehicles will pioneer the development of many components that later will become part of hydrogen-powered vehicles—gaseous fuel management systems and electric drive components, for example. In the long run, the experience gained from the hybrid electric or natural gas vehicles might well apply to hydrogen systems and actually accelerate their market entry.

Radical Innovation in Nonassembled Products

Significant innovations occur rarely in the lower left quadrant and tend to concentrate in the manufacturing process rather than the product. When these innovations change the business model as well as the technology, they tend to be brought to market by outsiders rather than industry incumbents. When the business model remains unchanged, insiders tend to dominate the innovation.

Consider electricity, for example—about as nonassembled as a commodity can get. Nuclear technology offered a radically different prime mover for electricity production when it was introduced in the 1950s. However, the technology was brought to market by industry incumbents, the large electric utilities and architect/engineer firms, and substituted directly for coal-fired generators. It did nothing to affect consumer behavior, and consumers could not distinguish nuclear-made electrons from those generated from coal. Hence, the incumbent business models were reinforced rather than attacked.

Now contrast nuclear as an innovation with the distributed generation of electricity. Providing electric service from larger numbers of small generators located close to the load can fundamentally change the business model from one dominated by regulated, vertically integrated electric utility companies to one dominated by user-generators. And characteristically, new companies with business models originating outside the utility industry are bringing distributed generation to market—Capstone Turbines or Northern Power Systems, for example.

In the case of hydrogen, these characteristic patterns seem to hold. Radical innovations in the business model—hydrogen production at the point of vehicle refueling, rather than remotely—are being proposed by new, entrepreneurial entrants to the vehicle fuel industry: Proton Energy Systems, for example. In contrast, industry incumbents tend to advance plans that look structurally similar to the current fuel infrastructure.

Incremental Innovation in Nonassembled Products

Finally, incremental process improvements in the form of continuous, small-scale innovations add significantly to the productivity of mature nonassembled products. There, innovation is almost exclusively the province of the incumbent producers who are motivated chiefly by cost competition. When combined with the leaps forward made by the few radical innovations, productivity gains tend to follow the pattern shown in Fig. 7-3. In glass manufacture and similar industries, the productivity gains that accrue from these incremental changes and those from radical changes in process architecture appear about evenly divided (Utterback, 1994). The hydrogen experience would probably be close to that shown in Fig. 7-3.

All these patterns simply distill accumulated observations of the way that technological change has occurred in other industries in other times. They are not laws of nature or even of economics. But they do suggest where entrepreneurial ventures are most likely to gain the most traction in any hydrogen transition.

The Entrepreneur as Agent of Change

A technology does not compete, and a gas does not innovate. For energy policy to succeed in creating a "new order of things," it must nurture the

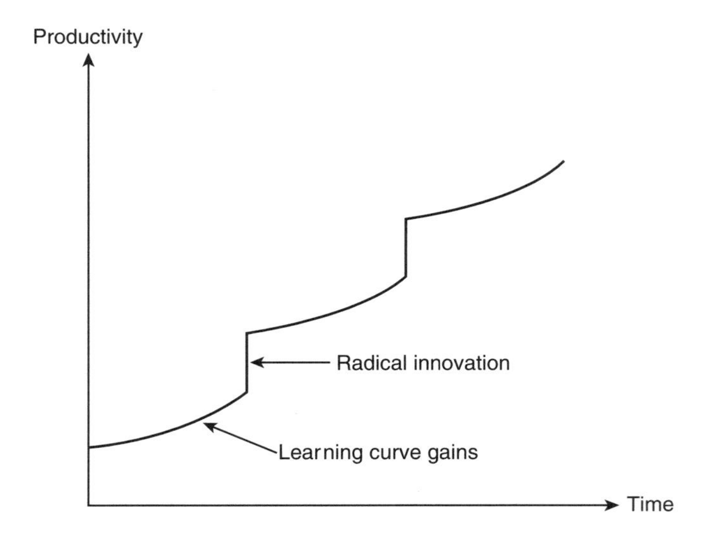

FIGURE 7-3. Innovation in nonassembled products.

real agents of economic change—entrepreneurs who can build economically sustainable ventures that grow into the mature hydrogen infrastructure. A National Academy of Engineering report (1995, 39) has noted:

> *The principal economic function of small entrepreneurial high-tech companies is to probe, explore, and sometimes develop the frontiers of the U.S. economy—products, services, technologies, markets—in search of unrecognized or otherwise ignored opportunities for economic growth and development.*

This suggests that policies intended to encourage entrepreneurship should be considered as one element of any transition to a hydrogen economy. Effective policy, however, requires a closer examination of how entrepreneurs create change.

Starting in Niche Markets: Small but Valued

Most radical innovations start in niche markets, communities of economic interest where the new technology solves some deeply felt problem. This is necessary because new technologies often serve customers poorly along the performance dimensions of value in mainstream markets. Radical innovations make the transition to the mainstream only after establishing a strong base in the niche. Mainstream businesses often find little of interest in these niches because they are too small or because they offer profit margins that are too narrow. Hence, niche markets are often left to the entrepreneurial venture. By contrast, incremental innovations usually fit well into mainstream markets, and market incumbents tend to develop and employ them better.

Consider microelectronics and the invention of the integrated circuit, for example. The transistor was invented in the early 1950s by an entrepreneurial start-up, Texas Instruments, despite generous research funding of the established electron tube companies by the Department of Defense. Though inferior to the electron tube in most consumer applications, the transistor solved an important defense need for lightweight, rugged electronics for radar, aircraft, and missile applications. Commercial applications followed much later, and even then had to begin in specialized niches like transistor radios where low cost, portability, and volume triumphed over sound quality in the minds of buyers.

But the rise of the transistor quickly revealed an even more urgent problem: large electronic systems were quite unreliable because each transistor was placed into the system as an individual component. No amount of quality control could overcome the reliability problems created by the manual interconnection of all these components. Once again, the problem was most severe in defense applications, and the Department of Defense was prepared to pay well for a solution. Once again, the industry leaders were well funded, and once again an outside company invented the answer—the integrated circuit, developed by Texas Instruments in 1958. The chief contribution of the military to the integrated circuit came from the assured market it provided, and not from its support of research by the industry leaders. The integrated circuit was not introduced into commercial markets until 1961 (Mowery and Rosenberg, 1998).

Much evidence suggests that these principles, derived from case studies across many industries over many years, might also apply in the hydrogen transition. For example, many incumbent auto companies are experimenting with fuel cell vehicles, admittedly with varying degrees of enthusiasm as they quite accurately judge that mainstream application is many years in the future. Without denying the appropriateness of this focus, a host of niche applications yet await the skilled entrepreneur—small, portable power systems for defense use, for example. And, as has happened so often in the past, successful technologies might emerge from these niche markets to stimulate consumer demand in a nascent hydrogen economy.

Enabling Technologies

Radical innovations rarely emerge into any kind of market, niche or mainstream, in isolation. Instead, their market penetration is assisted by innovations that we will term "enabling."

The PC, for example, succeeded in moving from its niche as a hobbyist's toy to mainstream commercial markets on the strength of two enabling innovations. The first was a microprocessor sufficiently powerful to run the software applications valued by consumers yet cheap enough to be sold to individuals through retail computer stores. The second was a set of software application packages containing functions that customers found valuable rather than merely amusing—principally word processing and spreadsheets (Bodde, 2004). Similarly, the integrated circuit drew upon diffusion and oxide-masking technologies that had originally been process innovations in the manufacture of silicon junction transistors (Mowery and Rosenberg, 1998). Such enabling technologies often provide the basis for the sharp productivity gains in nonassembled products, as illustrated in Fig. 7-3 (Utterback, 1994).

At the most aggregate level, innovations that improve the supply of hydrogen would be considered enabling by makers of the devices that use hydrogen, and devices that use hydrogen would be considered enabling by those who supply the hydrogen. But from the perspective of the new venture, complexities arise. For example, devices like fuel cell vehicles that create a demand for distributed sources of hydrogen might become powerful enablers for companies like Proton Energy who seek to offer such refueling systems. On the other hand, a fuel cell vehicle might be built around an entirely different refueling technology, say, hydrogen canisters manufactured centrally and sold in retail stores. In that case the innovation would remain enabling, but for somebody else.

Understanding the Business Model

Recognition of an opportunity becomes the point of departure for the entrepreneurial venture (Bodde, 2004). The most successful entrepreneurs build this opportunity into a complete business model embracing two core concepts:

- Value creation: an accurate understanding of how customer value can be created
- Value capture: a way for the new venture to build for itself an enduring competitive advantage in providing that value

Taken together, these concepts define for the entrepreneur and those who invest in new ventures the complete business model—a rational explanation of the present and future success of the enterprise. Developing a powerful business model presents a special challenge to entrepreneurs in the hydrogen transition because any business built around a component innovation requires assumptions about the performance of the other components in the system. For example, significant growth in the market for a new venture that offers, say, a hydrogen storage technology will depend upon acceptable performance of hydrogen-using devices like fuel cells.

Thus, we should think of the business model as a dynamic concept, continually adapting to the opportunities and threats in the business environment. These dynamics can be seen in Fig. 7-4. There, customer value finds its origins in a set of societal needs, which spring from the values, aspirations, and worries of the culture in which the new enterprise dwells. For the hydrogen transition, these needs include environmental protection and fuel security in addition to the more customary needs for personal mobility, which seems common to all humankind.

Societal need, unfortunately, does not always translate well into the market. Entrepreneurs can serve societal purposes only to the extent that those purposes also find expression in the marketplace as goods and services to be bought and sold. Personal mobility needs appear well understood in the marketplace and give rise to the fuels and vehicle infrastructure now in place. By contrast, environmental and energy security needs are only partially manifest. If entrepreneurs are to accelerate a transition to hydrogen, then finding a way to express these latent needs in the marketplace should become a principal goal of energy policy.

We can call the subset of societal needs that are recognized in the marketplace "customer value." This value might be explicit or it might be latent, unrecognized by most entrepreneurs and even unarticulated by customers. But above all, the value must be real, not contrived. Technology creates a powerful lever for building customer value. Indeed, technology that meshes well with the value that customers seek in the marketplace will always trump technology that is highly advanced.

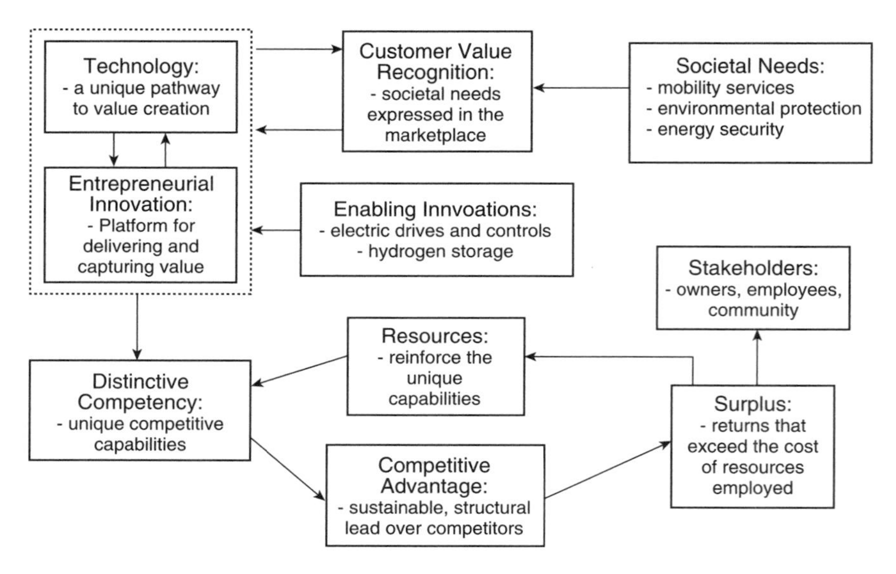

Adapted from David L. Bodde, *The Intentional Entrepreneur: Bringing Technology and Engineering to the Real New Economy*, M. E. Sharpe, Armonk, New York. Copyright 2004, M. E. Sharp. Used with permission

FIGURE 7-4. The dynamics of the business model.

Technology, however well matched to customer value, remains incomplete. In addition, entrepreneurs must build organizations capable of delivering the product or service and able to capture enough of the value to retain the interest of its owners, employees, and investors. The organization itself becomes an entrepreneurial innovation, the heart of the business model for the new company. The entrepreneurial innovation adds to the technology a delivery platform in the form of organizational capabilities, people, machines, and so forth. In most cases, either the technology or the entrepreneurial innovation or both benefit from enabling technologies. These are not a part of the new venture but are provided instead by the general business environment. For example, the online retailer Amazon did not invent the Internet but rather adopted this enabling innovation for its own purposes. In the case of the fuel cell vehicle, the electric drives and gaseous fuel controls developed for hybrid electric vehicles and natural gas vehicles serve this purpose.

As Fig. 7-4 shows, a combination of technology and entrepreneurial innovation that is truly unique to the new venture and unavailable to competitors becomes a distinctive competency. These competencies in turn create a sustainable competitive advantage. As a result of this advantage, the venture builds a surplus. Part of that surplus returns to the stakeholders of the enterprise: chiefly its founders, investors, employees, and community. The remainder becomes a resource to reinvest in nurturing the distinctive competency of the enterprise. The most powerful business models establish the virtuous cycle of Fig. 7-4, in which the success of the new venture becomes self-reinforcing—the better the company becomes, the better it can become.

Implications for Energy Policy

The experience of entrepreneurs and innovators in other fields and other times offers lessons for the design of energy policy today. Above all, it suggests that the cultivation of entrepreneurial behaviors and skills could materially assist in accelerating the pace of transition and achieving its public purpose. This acceleration might be achieved in the following ways.

Learning from the Market

No amount of clever analysis can help government planners know what transition pathway is right—any more than the central planners of the former Soviet Union could know which technology pathway would lead them to success in establishing a microelectronics industry, a goal that they never achieved. Rather, the competitive advantage of a market economy springs from its capacity for the empirical, and each new venture can be viewed as a reasoned economic experiment. In effect, the entrepreneurs become the "drosophilae" of the new energy economy. Their successes and failures provide important clues to the pace and direction of any incipient hydrogen transition. This suggests the first implication for energy policy—that policymakers should pay attention to the signals communicated by the private marketplace as its programs evolve through the transition.

Business Model Innovation

New technological advances are not always required for radical innovation. Consider Federal Express, for example, inventor of the hub-and-spoke model for overnight package delivery. Federal Express employed technologies that were no different from those used by any other competitor in the shipping business. But the logistic superiority of moving all packages to a single center for subsequent redistribution rather than sending them point to point soon gave Federal Express a decisive advantage. Competing firms either adopted this model or dropped out of the business.

This and similar cases suggest that policy thinking must not remain confined to the simple substitution of hydrogen technologies into the current business models for vehicles and fuels. Rather, entirely new business models might be required. For example, hydrogen might be produced in large-scale facilities and packaged onsite into portable fuel canisters. The fuel canisters could be sold at retail stores anywhere and loaded into the vehicle in much the same manner that a videocassette is loaded into the player–recorder. Alternatively, hydrogen might be manufactured at the point of loading into the vehicle, perhaps even in the home if adequate safety standards could be developed and their goals achieved. Thus, government technology development programs should remain alert to the emergence of new business models and nurture these.

Stimulating Entrepreneurship in Hydrogen

Entrepreneurs follow opportunity, usually seeking to establish an economically significant business within about 10 years. This allows the investors—who have exchanged a protracted period of illiquidity and high risk for the prospect of extraordinary gain—an opportunity to recover their cash from the enterprise, either through sale to a larger firm or a public stock offering. As a practical matter, twice that time is frequently required, but 10 years is a typical initial expectation.

Because entrepreneurs follow opportunity, public policies will find it difficult to promote a selected sector like hydrogen as distinct from the mainstream of opportunities. Perhaps the strongest instrument would be constancy of intent—a set of public priorities that are likely to stay in place for the 10-year planning horizon of the entrepreneurs and their investors. Venture capital investors know very well how to price risk, but they have little capacity to address policy uncertainty and simply walk away from it.

Other effective policies would seek to align public needs with private markets, allowing the societal preferences for clean air and fuel security to be expressed in customer preferences as well. Consider, for example, a mandatory cap on carbon emissions enforced by a requirement that suppliers and users of fossil fuels hold tradable rights for each ton of carbon they produced—the so-called "cap and trade" approach. This would have the effect of placing a price on carbon emissions that would eventually be included in the price of all goods and services. Technologies that are environmentally benign, like those based on hydrogen, would thus gain an advantage over those that are less benign. Taxes on the burning of high carbon fossil fuels would have the same effect.

Finally, hydrogen-specific investment programs might stimulate entrepreneurship. The government's Advanced Technology Program and the Small Business Innovation Research Program provide examples that might be applied specifically to hydrogen. In some cases, federal agencies have even opened their own venture capital operations—the In-Q-Tel partnership investing in companies related to the telecommunications needs of the Central Intelligence Agency, for example.

To be sure, public policy will always remain a secondary stimulant to entrepreneurship, relative to the primary rewards of the private market. But, even within this inherent limitation, improvements are possible and need to be made.

References

Bodde, David L. 2004. *The Intentional Entrepreneur: Bringing Technology and Engineering to the Real New Economy.* Armonk, NY: M. E. Sharpe.

Energy Information Administration. 2001a. *Annual Energy Review, 2001*. Washington, DC: U.S. Department of Energy.

Energy Information Administration. 2001b. *A Primer on Gasoline Prices, July 2001*. Washington, DC: U.S. Department of Energy.

Mowery, David C. and Nathan Rosenberg. 1998. *Paths of Innovation: Technological Change in 20th Century America*. Cambridge, UK: Cambridge University Press.

National Academy of Engineering. 1995. *Risk & Innovation: The Role and Importance of Small High-Tech Companies in the U.S. Economy.* Washington, DC: National Academies Press.

National Research Council (NRC). 2002. *Effectiveness and Impact of Corporate Average Fuel Economy (CAFE) Standards*. Washington, DC: National Academies Press.

Utterback, James M. 1994. *Mastering the Dynamics of Innovation*. Boston, MA: Harvard Business School Press.

CHAPTER 8

Hydrogen from Electrolysis

Chip Schroeder

Electrolysis of water is a method used to produce hydrogen. It is a viable alternative to the common practice of reforming hydrocarbon fossil fuels for today's industrial gas markets and for tomorrow's energy markets. My vantage point is that of the CEO and co-founder of Proton Energy Systems, a relatively new company that has developed a unique electrolysis technology. I will not be promoting our company in this chapter, but I will use our experience to explain the practical issues and opportunities confronting the several companies that are today working to advance various electrolysis technologies.

In a nutshell, many companies, including Proton, are pursuing electrolysis technologies and products for a few simple reasons. First, when derived from water, which has no carbon impurities, electrolytic hydrogen is chemically perfect for fuel cells. There is in fact no better fuel for fuel cells. Second, electrolysis is the only practical means of making fuel from renewable energy resources. Third, electrolysis technologies are relatively mature; they are the basis for products that are already used commercially in today's industrial gas markets. The fundamental message I hope to deliver in this chapter is that the closer you look at electrolysis, the more interesting it becomes as a practical answer to the question: Where will the hydrogen for fuel cells and the broader hydrogen economy come from?

Background on Proton Energy Systems

Proton Energy Systems, Inc. received a first round of venture capital funding in 1996. Several funding rounds later, the company has grown from its original 5 employees to 115. Our business plan was and still is to be the leader in making products that harness proton exchange membrane (PEM) technology to serve high value commercial markets. Our core competency is in PEM electrolyzers. We make cell stacks and incorporate those proprietary modules into working systems.

Proton's first products are designed to provide onsite hydrogen generation capability in industrial markets that now obtain hydrogen in cylinders via costly truck delivery. The same products, made on a larger scale and properly sited at fueling stations or perhaps even homes one day, can be the source of onsite hydrogen for meeting vehicle fueling needs. We are also developing products that "marry" our core electrolysis hydrogen-producing components with fuel cells to create energy storage and power-quality products. In effect, our hydrogen generation technology enables fuel cells to serve a wider array of power-related applications, some of which will probably prove commercially viable well before the fuel cell car. Figure 8-1 shows the leading technology pathways connecting hydrogen electrolysis to the end use market.

Proton is a public company with about $100 million of cash on hand, but we know that the challenge of introducing disruptive technology to any market, including the automotive market, which is the largest manufactured product market in the world, will take huge amounts of capital. Accordingly, Proton and the other smaller independent companies working on hydrogen and fuel cells share a common need to identify and capitalize on the high value niche markets that can produce revenue streams today. The emphasis of the discussion and exhibits that follow is to explain

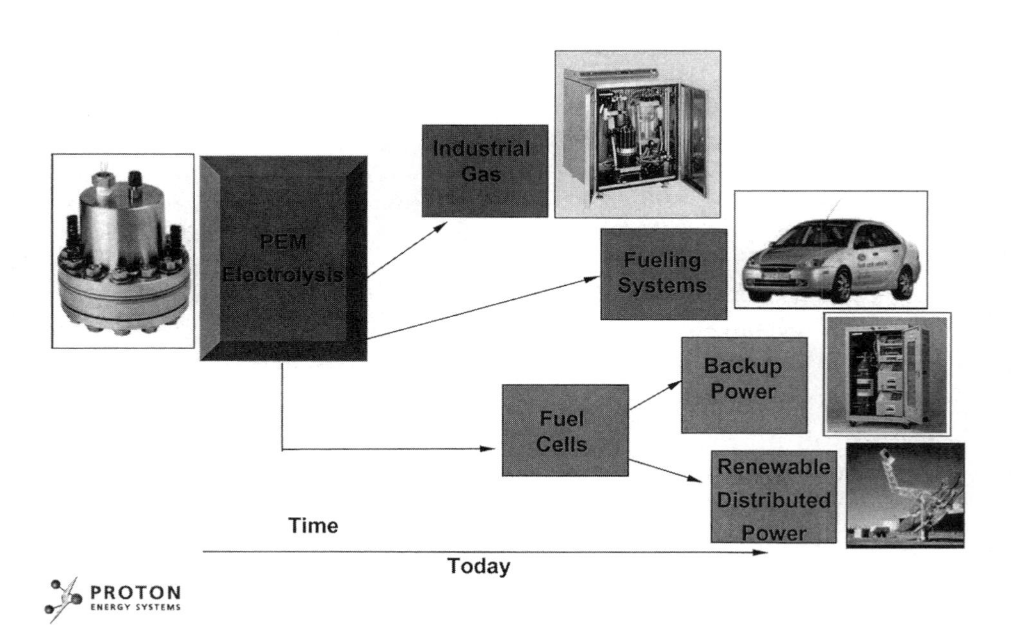

FIGURE 8-1. Electrolysis technology pathways.

those markets and the important role that electrolysis will play in those markets.

PEM Electrolysis

PEM electrolysis has been used successfully for nearly three decades in submarines and in spacecraft to generate oxygen for human life support needs. Several of our founders came out of United Technologies, Inc., which has been the leader in PEM electrolysis for these high value defense-related applications. We make and sell several sizes of PEM electrolyzers that are literally fuel cells in reverse: Water and electricity are the inputs, and hydrogen and oxygen are the products. Our systems convert electric energy into chemical energy, whereas fuel cells convert the chemical energy of hydrogen into electric energy. PEM electrolyzers incorporate a solid polymer membrane that helps manage the electrolysis process in such a way that hydrogen ends up on one side of the membrane, while oxygen remains behind, suspended in the water that serves as the "feedstock" for our systems. So we produce pure hydrogen and, if we need it, pure oxygen.

Why would any self-respecting engineer want to make a fuel cell that runs backward? After all, if the excitement surrounding fuel cells is that they can cleanly and efficiently convert hydrogen into electricity, why squander that electricity by turning it back into hydrogen? From a "net energy" perspective, doesn't it take more British thermal units (BTUs) of electricity than are contained in the hydrogen produced from electrolysis?

The answer begins with an acknowledgment that the amount of energy consumed in PEM electrolysis is indeed greater than the amount of energy in the resulting hydrogen. But this tradeoff can, in many cases, make good economic sense. For example, if the electricity used to make electrolytic hydrogen comes from low priced coal or nuclear power sources, and if the hydrogen is then used to replace high priced fuels such as gasoline, we have effectively transformed coal or nuclear resources into transport fuel. In such practical applications, the economic value added overwhelms the net energy loss. An even more compelling justification for electrolysis comes from the desire to see renewable power make an impact on transportation markets. Renewable resources, such as wind, solar, or geothermal power, give us electricity, but not fuel. The only practical way to turn renewably generated power into fuel is through electrolysis.

Another key to the surprisingly good economic basis for electrolytic hydrogen is that the hydrogen can be made onsite, that is, at or near the point of end use, thereby minimizing or eliminating transport costs. In effect, electrolysis takes advantage of the existing infrastructures for electricity and water. The cost of an electrolyzer sited at a gas station and sized to fill 10 to 20 cars per day is far less than the total capital cost of a new large scale steam-methane reformer that requires a new pipeline or truck-based delivery infrastructure. For this reason, various experts have concluded that electrolysis will have a role in the introductory stages of the hydrogen fueling marketplace. The longer-term role of electrolysis for fueling will depend upon how the economics of converting electricity to hydrogen compare with the economics of other fueling options.

The economics of electrolysis are helped by the surprisingly good electrochemical efficiency of this technology. As Fig. 8-2 illustrates, PEM electrolyzer stacks exhibit an inverse relationship between efficiency and current density, or amperes per square foot. When low levels of current are applied to the stack, resulting in lower output of hydrogen, the efficiency of the process can exceed 85 percent. That is, more than 85 percent of the BTUs of electric energy is converted to BTUs of hydrogen chemical energy.

Much like an internal combustion engine, a PEM stack gets less efficient the harder it is "driven." Our systems today confront a trade-off between efficiency and capital cost. The stacks in our commercial systems operate at below 80 percent efficiency because the PEM cells are expensive. As the cost of cells and cell stacks comes down, we will be able to put more cells into each stack (with correspondingly lower current density per cell) and higher resulting efficiencies.

Relatively Mature Technology

- Electrolysis is 85% efficient

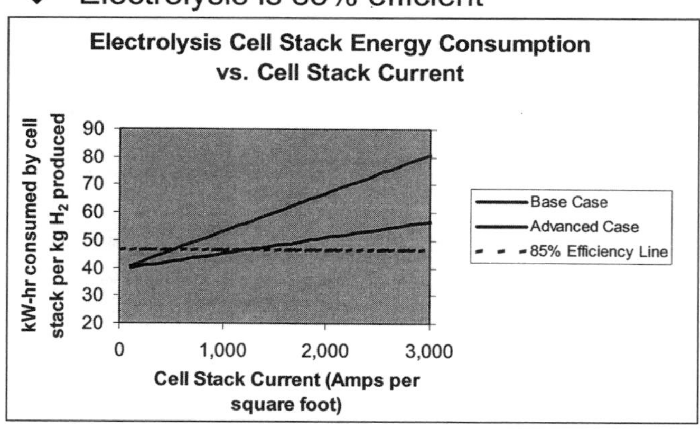

- Pressure feature
- Temperature
- System design

FIGURE 8-2. Efficiencies of hydrogen electrolysis.

The math for translating electricity into hydrogen-based fuel cell transport is fairly straightforward. The theoretical efficiency of converting electricity into hydrogen via electrolysis is 39.4 kilowatt-hours per kilogram (kWh/kg) of hydrogen. Assuming we place a 75 percent efficient electrolyzer system at a typical gas station, the electricity requirement of hydrogen rises to 39.4 ÷ 0.75, or 52.5 kWh/kg.

If we use that hydrogen to fuel a current generation fuel cell demonstration vehicle, that vehicle will travel 90 to 100 kilometers (km) (or 55 to 60 miles) on 1 kg of hydrogen. So a kilogram of hydrogen "costs" 52.5 kWh to produce and provides better than 55 miles of driving, or just about 1 kWh of electricity to drive 1 mile. If the cost of electricity at the gas station is, say, 7 cents per kWh as assumed in Fig. 8-3, this equates to 7 cents per mile as the fuel cost of driving a fuel cell vehicle. That cost is perfectly competitive with today's gasoline internal engine automobile. If gasoline costs $1.70 per gallon, then a 20 mile-per-gallon car has a fuel cost of 8.5 cents per mile.

Most analysts are quite surprised when they first work through the economics of hydrogen fuel from electrolysis. The presumption is that the net energy cost of making hydrogen from electricity is prohibitively high. How can the fuel value at the gas station possibly be greater than the fuel value that went into making electricity in the first place? The answer, of course, is that the cost of the BTUs used to make the electricity is much lower than the value of transport fuel. The variable cost of electricity, including fuel, operation, and maintenance, at a coal fired generating plant

The Economics Actually Work

Current electrolysis technology, *and*

Current fuel cell demonstration vehicles

Get *1 mile per kWh*

Bottom line: 7-cent electricity competes with gasoline at $1.50 per gallon

FIGURE 8-3. The economics of electrolytic hydrogen as a motor fuel.

is only about $0.01 per kWh, or about 15 to 20 percent of typical commercial electric prices. Again, on a gasoline equivalent basis, the generating cost of base load electricity is perhaps one-eighth the value of the fuel that it can replace if electrolyzed and used in a fuel cell vehicle. It's as if we start with a gallon of water at the utility generator, but when it gets to the gas station the water has turned into wine. Sure, we spilled some, but wine is worth enough more than water to overcome the shrinkage.

So the reality is that the variable cost of fueling a fuel cell vehicle with hydrogen from water is much more interesting than most people initially anticipate. Moreover, electrolysis technology is modular and scalable. As a result of all these factors, hydrogen from electrolysis is gaining credibility as perhaps the most logical way to achieve the introductory phase of the hydrogen fueling infrastructure.

One final issue to consider is the effect on utility economics if use of the utility grid increases in the face of a growing electrolytic hydrogen market as a motor fuel. Generating capacity and wires that are not fully utilized during off-peak periods can now be effectively channeled to meet transportation fuel needs. Capacity factors thus improve, and rates charged to fueling stations may drop. Couple this with the inevitable political interest that will derive if utility rate making and practices become intertwined with retail transportation fuel costs, and the implications for electrolysis as a source of fuel get ever more intriguing.

Building on Today's Industrial Hydrogen Products

Electrolyzers used for future fuel cell vehicle fueling applications are straight derivatives of electrolyzers used today for making industrial hydrogen. Proton, Stuart Energy, and several other European and Japanese companies are building and selling electrolyzers for industrial needs. The value of hydrogen in distributed, relatively small volume industrial markets is far greater than the value of hydrogen based just on its energy content. This is because industrial users of hydrogen often are taking advantage of the unique chemical value of hydrogen, not its energy value.

To understand this point requires a brief examination of the structure and nature of the industrial hydrogen marketplace. Virtually all the hydrogen used in today's industrial gas markets comes from large scale natural gas processing plants, called reformers. These plants, owned by major industrial gas companies like Air Products, Air Liquide, Praxair, Linde, and BOC, use steam and catalytic processes to break the natural gas molecule into hydrogen and various carbon-based elements, including carbon dioxide. Large steam-methane reformers are usually located near a major hydrogen-consuming petrochemical facility such as a refinery. The process of hydrocracking—adding hydrogen to hydrocarbon molecules to improve their fuel properties—is a major market for hydrogen. But hydrogen is also used to make a vast array of chemicals as well.

The price of hydrogen paid by these large users tends to be directly related to the cost of natural gas. But in the industrial markets for hydrogen, particularly in high tech manufacturing, where the volumes of hydrogen are smaller, the logistical costs of delivering and storing hydrogen become the dominant cost factors.

A portion of the hydrogen produced at large steam-methane reformers is targeted for so-called merchant markets. Merchant customers can shop among industrial gas suppliers, typically purchasing on a several-year contract basis. Their hydrogen generally gets delivered by truck. Merchant customers typically use hydrogen in high purity manufacturing processes. Manufacturing companies like semiconductor makers, specialty metal manufacturers including titanium products, glass manufacturers, and a variety of other high technology manufacturers use hydrogen to prevent oxidation.

Semiconductors, as an example of a merchant hydrogen market, are made in high temperature fabricating ovens, or "fabs." In the fab, metals at very high temperatures are vapor deposited onto the silicon chip. If any oxygen is present in the fab, the metals can get oxidized and the chip would be ruined. The role of hydrogen in the process atmosphere in chip making is to intercept oxygen and neutralize it into water before the oxygen damages the materials being processed. The exact same principle is at work in glass or high temperature metal fabrication, where oxygen creates impurities and structural weakness.

Clearly, hydrogen is a critical, high-value element in such manufacturing environments. These are niche markets when compared with potential hydrogen energy markets. But these niche markets amount to several hundred million dollars per year and represent true commercial markets for Proton and our competitors. Proton today makes and sells electrolyzers that produce about 40 standard cubic feet (SCF) of hydrogen per hour, or about one kilogram per day. The selling price of such a unit is about $50,000. These units consume about 6 kW of electricity, so the value of our equipment in these settings is about $8000 per kW. Compare this figure with the price value of fuel cells for automotive power production, which is widely recognized to be in the $100 per kW range. Clearly, the value of PEM technology in the distributed industrial hydrogen marketplace is far higher than in the transportation energy marketplace. Proton's business plan is to capture these high-value-chemical based markets today, while advancing research and development into cost reductions that will enable this technology to become competitive in the emerging vehicle fueling markets of the future.

Power Quality: Hydrogen as a Battery Alternative

Secondary or backup power is often worth far more than primary power. For example, the value of grid electric capacity is roughly $1000 per kW. That is, if one takes the depreciated cost of an average generating plant, a transmission plant, and a distribution plant, a typical customer is paying capacity charges that support about $1000 per peak kilowatt consumed. The figure may be somewhat higher for residential customers and lower for large industrial customers, but $1000 is a reasonable average.

The value of backup power is as much as three to four times the value of primary power on a kilowatt basis. For example, the lifecycle cost of the backup power systems found at the base of a cell tower, which now consists of a bank of lead acid batteries and a diesel or natural gas fired combustion engine, is between $3000 and $4000 per kW. Critical power facilities for data processing centers and the like are also in this cost range. The simple fact is that customers need electricity and will pay a considerable insurance premium to obtain assurance of uninterruptible power. In the case of cellular phone service providers, their federal FCC license may be at risk if they are unable to demonstrate adequate operating capability in the event of grid outages.

Electrolyzers coupled with fuel cells are an alternative to battery-based backup power. The system is actually quite simple, consisting of four components. First, an electrolyzer operating off grid power makes hydrogen. Second, a conventional hydrogen storage tank stores the hydrogen. The tank is designed to hold enough hydrogen to meet the ride-through requirements of the power user. It could be sized to hold hydrogen for hours or even days of backup power, whereas a battery operates in discharge mode for only an hour or two at most. Third, a fuel cell instantaneously comes to life when power controls sense that the grid has failed. Finally, integrating hardware and software makes the entire system work.

The size of the fuel cell is what determines the power output of the system. The electrolyzer, on the other hand, can typically be considerably smaller than the fuel cell. Because the grid is 99.5 percent reliable, it makes sense to design the system to make hydrogen much more slowly than it uses it to produce power, so a small "trickle charging" electrolyzer makes sense. It should be apparent that if the system can sell for $3000 to $4000 per kW, the value of the fuel cell component for such a system should be $1000 or more per kilowatt. Accordingly, the value of fuel cells for such high value stationary applications will be 10 times their value for automotive duty. Therefore, long before fuel cells are operating commercially under the hood of a car, they will be operating in tandem with electrolyzers in power quality applications.

Figure 8-4 helps explain the value of PEM technology for each of the applications we have been discussing. First, PEM technology embodied in electrolyzers has a value in the $5000 to $8000/kW range when deployed to produce industrial hydrogen for distributed merchant markets. To the right, fuel cells for automotive use will need to compete with internal combustion engines that cost $50 per kW. Taking into account the higher efficiency of fuel cells, analysts put the value of fuel cells for vehicles at less than $100 per kW. This is the cost target of every major fuel cell developer, including GM,

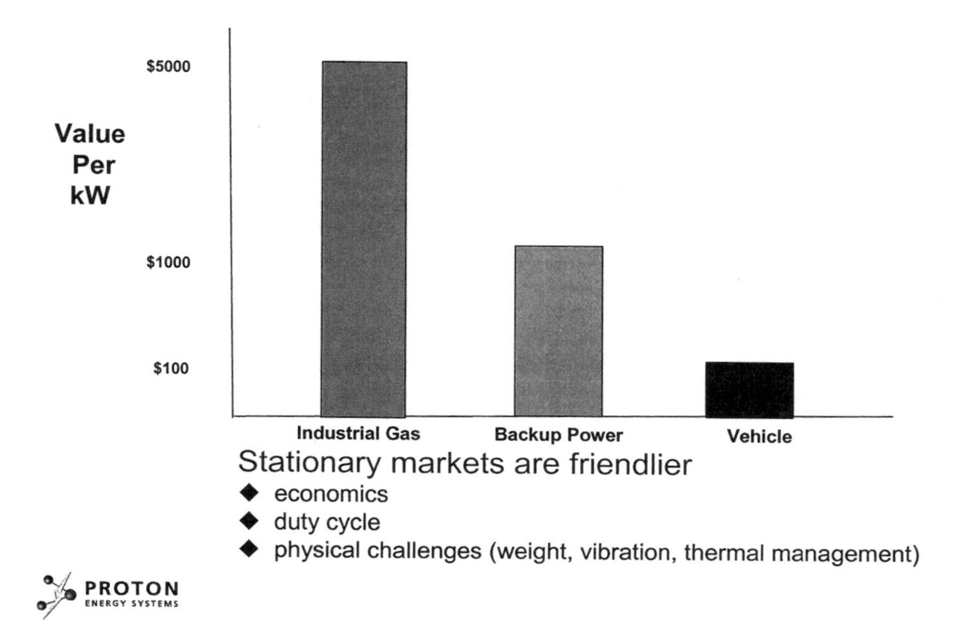

FIGURE 8-4. Transport vs. stationary market values.

Ballard, UTC Fuel Cells, Honda, and Toyota. It should be apparent why companies like GM are now indicating that they will be deploying their fuel cells in stationary power quality markets as interim tests and applications of their fuel cell technology.

Fuel from Renewable Resources

When we founded Proton, the ultimate vision for the company was to find a way to utilize PEM electrolysis to make hydrogen from renewable power—wind, solar, geothermal, and hydro. We all recognize the importance of nondepleting and nonpolluting renewable technologies for creating a sustainable energy marketplace. But renewable resources by themselves have a hard time affecting our most troubled energy market—the transportation sector. The value of renewable resources is currently confined to the electricity sector, while the sector that contributes most to energy depletion and import vulnerability is the transportation sector.

This is where electrolysis holds enormous promise. Electrolysis presents a path to hydrogen production from renewably generated electric power. From an energy perspective, electrolysis is literally a way to transform electricity into fuel. Electrolysis is thus the means of linking renewably generated power to transport fuel markets.

Many fuel cell vehicle demonstration programs taking shape around the world reflect a recognition that the energy for fuel cell vehicles can and should come from renewable resources. Fuel cell vehicles are nonpolluting in themselves, but if the hydrogen that fuels these vehicles comes from reformed natural gas, the pollution and depletion cycles have not been broken. Electrolysis coupled with renewable resources offers the only practical means of making transport fuel with no pollution or depletion anywhere along the path.

Electrolyzers are part of a number of European fueling programs, including several fuel cell bus demonstrations. The U.S. Department of Energy has recently taken a somewhat proactive view in support of electrolysis for fueling. It has encouraged the various teams working on hydrogen fueling infrastructure projects to incorporate an electrolysis component.

Proton has been involved in a number of successful demonstration projects for making hydrogen from renewable resources, including solar thermal, solar photovoltaic, and wind. Figure 8-5 shows one of our

FIGURE 8-5. Solar thermal hydrogen and electricity generator.

electrolyzers operating in tandem with a 25-kW solar thermal generator. During daylight hours, a portion of the electricity produced by the solar unit is fed to our electrolyzer. The electrolyzer makes hydrogen that can be stored in simple propane tanks and then fed back to the thermal combustion unit to make power on demand from inherently intermittent renewable power.

This type of project has application either as grid support in developed electricity markets or as stand-alone small power systems for off-grid and developing-country applications. Looking to the future, projects like this can serve not only as local electric utilities that can operate independently from central generation and transmission, but whose stored hydrogen can serve as a source of transport fuel.

Electrolysis as "Energy Arbitrage"

I have argued thus far that electrolysis can add important new dimensions to energy markets. In thinking about these dimensions, let's begin with the fact that hydrogen is not primary energy. It takes primary energy sources, such as hydrocarbons, or electricity to make hydrogen. While hydrogen can be burned as a replacement for hydrocarbon fuels, its value is likely to be greatest as a fuelstock for fuel cells with electricity as the output.

So electrolysis is really a process enabling "arbitrage" among various primary energy sources. There are three fundamental ways in which this arbitrage is or may be of commercial significance, as shown in Figure 8-6.

The first is for on-peak versus off-peak, or intertemporal, electricity price arbitrage. If the difference between the prices of power at different times gets wide enough, it makes sense to convert and store the cheap electricity in the form of hydrogen fuel for subsequent reconversion into electricity.

Second is the arbitrage value of primary versus uninterruptible electricity. Some electricity users value electric service far more than its cost. If the value difference between primary and uninterruptible power becomes great enough, it again makes sense to convert and store the cheap primary electricity in the form of hydrogen fuel to ensure the continuity of high value power.

Finally, there is crossover energy arbitrage. If the cost of a hydrocarbon fuel gets high enough in relation to the cost of electricity, hydrogen via electrolysis may be a lower cost energy substitute for the hydrocarbon. This involves the concept of using a lower grade of hydrocarbons, such as coal, to make the electricity for electrolytic hydrogen production that, in turn, can substitute for much higher grade hydrocarbons in a downstream application.

All the commercial or potentially commercial energy-related applications for electrolytic hydrogen fall into one or more of these three categories. For example, vehicle fueling makes sense because of crossover arbitrage.

Electrolysis: Fuel From Electricity

Interesting "Arbitrages"

1. Time of day differentials

   2:1 / price difference required

2. Quality differentials

   grid power cost: $1,000 / kW

   backup power value: $4,000 / kW

3. Fuels "crossover"

   low cost to high value BTU's

   (e.g.,coal to gasoline)

FIGURE 8-6. Energy arbitrages from electrolytic hydrogen.

- Transform renewable power into hydrogen
- Hydrogen can provide
  - power on demand
  - transport fuel
- The critical link between renewables and transport markets

FIGURE 8-7. Sustainable power from renewable sources.

The value of coal is far less than the value of gasoline. Indeed, coal at $20 per ton costs about $0.82 per million Btu, whereas gasoline at $1.75 per gallon works out to $15.40 per million Btu. The energy losses of electrolysis are such that the huge cost differentials among fuels support the idea of crossover arbitrage.

Similarly, the value of electricity for different needs creates a quality arbitrage. And finally, because there are no practical bulk electricity storage options other than pumped hydropower, electrolysis and hydrogen conversion promise to play an invaluable role in enabling power on demand from inherently intermittent renewable resources.

Figure 8-7 is arguably the most important. It reminds us that electrolyzers and fuel cells can create the path to energy sustainability. By converting electricity into fuel, electrolysis lets us make use of electricity in entirely new ways. The hydrogen we can now make from renewable power can be stored for subsequent use during periods of critical demand. Nighttime power on demand from solar photovoltaics will become a reality with electrolytic hydrogen, as will power from wind resources even during periods of calm.

Most important of all, hydrogen from electrolysis links renewable resources to transportation energy markets in a practical way. It is hard to imagine any more valuable contribution to the world's energy needs than to create the path to nonpolluting and sustainable fuels. The promise of electrolysis is just that huge. Very real challenges lie ahead, in cost reduction, commercial acceptance, and policy conformance. But the rewards surely merit the challenge.

CHAPTER 9

The President's U.S. Hydrogen Initiative

Steve Chalk and Lauren Inouye

Since the 2003 State of the Union address by President G.W. Bush, the media and general public have voiced their interest in the potential of hydrogen to meet the energy needs of the United States (U.S.). It is encouraging to see popular interest in energy issues, and to see enthusiasm from the private sector and the U.S. public regarding the potential of hydrogen and fuel cell technologies. At the same time, it is important that coverage of hydrogen and fuel cells be balanced, and that people also be aware of the full spectrum of Department of Energy (DOE) activities.

The president's Hydrogen Initiative should be viewed in terms of what the program is, rather than what it is not. Together, the president's Hydrogen Initiative and FreedomCAR programs plan to invest $1.7 billion in federal funds over 5 years to develop a more sustainable energy and transport system. However, the program has been mislabeled as a panacea to the world's sustainable development challenges, and this is not the case. Hydrogen is not the solution to suburban sprawl, to inadequate transportation in the developing world, or to all pollution in the United States. It is, however, one of the most viable energy options for transportation that allows for energy independence and has the ability to virtually eliminate greenhouse gases and air pollutants.

It is important that hydrogen not be "oversold" to the public, leading to unrealistic expectations about either the time frame of hydrogen fuel cell vehicles (FCVs) or the scope of problems they can help address. Such misunderstandings may result in disappointment when technologies are not available or backlash from individuals who expect more than any individual technology can promise. Hydrogen FCVs are a promising technology, and one that has tremendous potential to address transportation and energy problems, but they are only one element in the larger U.S. strategy to secure a healthy environment and sound economy for future generations.

Overview

Hydrogen has the potential to be an important component of the long term solution to U.S. energy needs. The DOE research timeframe aims to achieve enough significant technological progress to allow the private sector to make a commercialization decision in 2015. To achieve this goal, the DOE has developed a Hydrogen Posture Plan which envisions long term goals and technology milestones through 2015. In addition, the DOE Office of Energy Efficiency and Renewable Energy has laid out the more detailed Multiyear Research, Development and Demonstration Plan, which has specific technology targets and evaluation points through 2010. As the technology develops, the DOE will consider options and continuously evaluate the costs and benefits of various methods of hydrogen production, delivery, and storage. The program will evolve and adapt as research activities are completed, and as the DOE continues to weigh technological options, environmental impacts, and economic trade-offs. We anticipate a transition period of several decades as technologies mature and conventional internal combustion engine cars are phased out in favor of superior low emissions gasoline or diesel vehicles, high efficiency hybrid electric vehicles, and then hydrogen vehicles.

Timing of the Hydrogen Transition

Some have argued for a more rapid transition to a hydrogen economy. However, the DOE does not mandate what vehicles individual consumers drive. Individual buyers will make that decision themselves. Unlike the Apollo Program, with which the president's Hydrogen Fuel Initiative has sometimes been compared, the government is not the ultimate customer of the technology we are helping to develop. In order for hydrogen vehicles to be widely adopted, businesses must offer vehicles that meet consumer requirements. However, financial risks involved in the development of hydrogen technologies and infrastructure are significant.

Without additional research demonstrating their promise, the business case for investment in the hydrogen economy is uncertain. Therefore, the DOE will support research and development (R&D) to help establish a business case and to enable a commercialization decision. Given the tremendous technical hurdles that must be overcome today to meet customer requirements, 2015 is an ambitious but achievable time frame.

Figure 9-1 shows the four phases in a transition to a hydrogen transportation energy economy in the United States. Wide adoption of fuel cell vehicles will most likely not take place before 2020, and a minimum of 20 years will be required for the majority of petroleum based internal combustion engines (ICEs) to be effectively replaced. The DOE's time line takes technological unpredictability into account but seeks to make a commercialization decision on fuel cell vehicles possible by 2015.

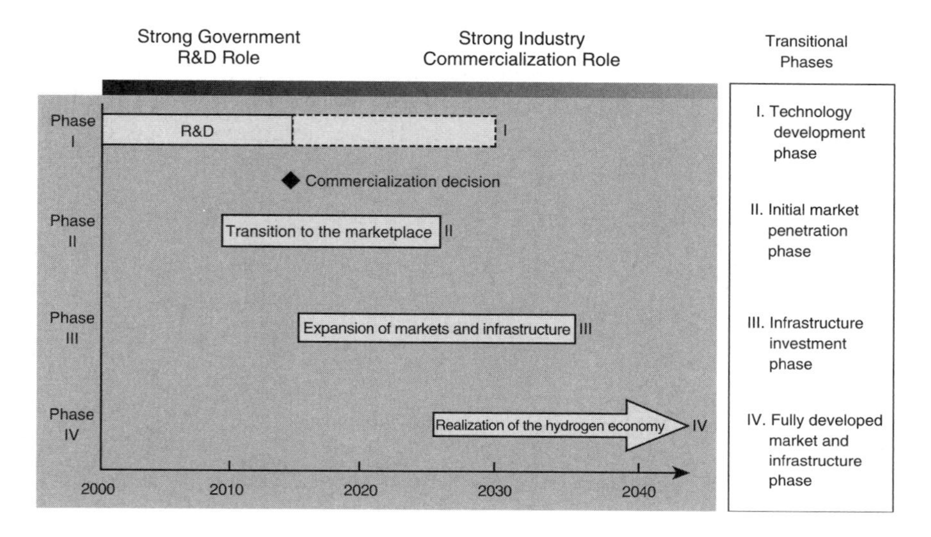

FIGURE 9-1. The hydrogen transition in the United States.

Our time frame and budget allow us to support the exploration of multiple production, delivery, and storage pathways, and to reevaluate the feasibility of different options with phased "go/no-go" decision points. Systems integration is crucial. Hydrogen production, storage, and delivery methods must complement each other, and the feasibility of one method will depend upon its compatibility with other parts of the system. This is why we have established a systems integration plan to analyze all models and pathways and to appropriately integrate individual efforts with overall objectives at the macro-system level.

The DOE time frame balances a desire to begin work on fundamental technical barriers to the hydrogen economy with a measured plan and methodology for systems analysis. Compressing our research schedule would not allow for the necessary learning and experimentation cycles required in R&D programs. Overly aggressive funding of research that is not well conceived could result in wasted taxpayer money.

For example, the DOE is currently funding research into 10,000 pound per square inch storage tanks. Storage and delivery are particularly challenging aspects of the hydrogen system and clearly need further research. However, if further research determines that low pressure, solid state storage is a better technical option, it would change our thinking about delivery and refueling infrastructure. Careful coordination of the development of all components of hydrogen production, delivery, storage, end-use conversion, and infrastructure is needed.

In particular, the DOE Hydrogen Program emphasizes developing hydrogen infrastructure in parallel with hydrogen fuel cell vehicle technologies.

If one element of the hydrogen system is not able to meet customer requirements, it is unlikely that the system as a whole will be able to function. For example, having the infrastructure to enable every driver to refuel their hydrogen fuel cell vehicle near their home or place of business is crucial to the success of the hydrogen transition. There is a need to coordinate elements of the hydrogen value chain, so one component does not delay progress on another.

Countering those who push for a faster hydrogen transition, others argue that the DOE is rushing the hydrogen transition and want to postpone hydrogen-related research. Often, their proposed alternatives focus on near term policies and regulations to address pollutants, greenhouse gas emissions, and oil dependency. However, substantial efforts to improve near term energy efficiency and environmental impact do already exist. The DOE supports a broad and robust portfolio of activities in energy efficiency, renewable energy, nuclear energy, fossil fuels, carbon sequestration, and other research at the same time we support research on hydrogen and fuel cells. The two approaches are not mutually exclusive. Many short term, immediate actions supported by the DOE will make a greater near term difference to current fuel consumption and air quality than the hydrogen program will, simply because FCVs will probably not be widely available until 2020. While resource constraints make it impossible to support every project, the DOE's work recognizes that it is important that both long term and short term solutions be supported.

Far from being mutually exclusive, near term efforts can actually support long term solutions. This is the case with hybrid electric vehicles (hybrids), which can meet customer requirements on performance, design, and durability, and which are commercially available today. More hybrids on the road mean greater fleet efficiency and cleaner air. The Bush administration supports an efficiency based income tax deduction for the purchase of new hybrids. In addition, through the FreedomCAR program, the DOE continues to support development of advanced hybrid components and research on electric power-train technologies. This research, which today helps improve hybrids of battery and ICE drivetrains, will be used in the future on the electric drive systems in FCVs. Other DOE research contracts sponsor work to improve energy storage and develop advanced lightweight materials. These technologies improve the efficiency, cost, and durability of hybrids, while also promoting development of FCVs.

Distributed production of hydrogen from natural gas is another example of a near term solution that helps reduce pollution, while building the infrastructure for renewable hydrogen production methods that are still in development.

It is important to recognize that short term measures, while essential, will not bring a total solution. While the Bush administration is supporting measures such as hybrid vehicle tax credits, ethanol production, and revisions in fuel efficiency standards, these interim strategies will not ultimately reverse our dependence on foreign oil or address environmental problems as hydrogen can.

Thus far, hydrogen is one of the few fuel options that gives the U.S. the option of a domestically based, zero emissions energy economy. Battery electric vehicles show similar potential, but without a significant breakthrough in battery technologies, they are unable to meet customer requirements. The DOE continues to support battery research, which may also be an important component of commercially viable FCVs, but hydrogen fuel cells appear to be a more attractive option at this time, and it would be foolish not to begin exploring their potential. By investing critical funding in R&D today, widespread commercial adoption of hydrogen fuel cell vehicles could be near completion in 2040 or 2050. Without the commitment and funding of the President's Hydrogen Fuel Initiative, we believe it is unlikely that hydrogen and fuel cells will be viable technology options when the interim measures viable today are no longer sufficient.

Other near term policies may also be part of the solution to energy security and environmental concerns. However, many policies in this arena are quite contentious and may have consequences for the economy and the competitiveness of U.S. businesses. It is risky to call for legislation requiring technical solutions when the needed technologies do not yet exist. In addition, it can take years for consensus to emerge around a sound policy, and even then, it is often difficult to implement the laws to ensure it is effective. Given the tremendous uncertainty of overcoming all the technical hurdles to a hydrogen economy, it is best to focus today's efforts on R&D and technology validation. In reaching our technical performance and cost targets, the DOE will enable tomorrow's policymakers to debate about a more proven technology. In the interim, the DOE will help support the further development and adoption of existing technologies, such as advanced combustion engines, hybrids, and renewable energy.

DOE's R&D Strategy

The DOE role is to stimulate and support early, high risk research with the most significant potential payback. The DOE, with its partners, will also test these technologies in real-world conditions through demonstration projects, and the information gathered will be fed back into research programs to help determine where problems and priorities lie. The DOE and its partners have identified the three greatest challenges to a hydrogen transition to be improving hydrogen storage, reducing the cost of hydrogen production and delivery, and making fuel cells less expensive and more durable.

Some FCV enthusiasts advocate policies and incentive programs to encourage widespread adoption of hydrogen fuel cell vehicles today, either among fleets or individuals. It is true that economies of scale in the automotive industry make it possible to produce technologies in mass production at a lower cost than the same product made in smaller quantities. However, while this is an appropriate strategy for some technologies, such as hybrids, we believe it is premature for hydrogen FCVs. It is far too early to bring hydrogen technologies and fuel cells down the cost curve with such policies. Today, FCVs do not meet customer needs and are not close to the cost targets for broad commercialization. It would be a waste of taxpayer funds to promote technology that did not meet customer needs, and could even backfire by popularizing negative perceptions about FCV performance or durability. In addition, the U.S. lacks the infrastructure, codes, and standards necessary to make any vehicle early adopter program involving the general public a feasible scenario.

When the hydrogen and FCV technologies are close to meeting customer requirements, and the business case is nearly established, the DOE will consider policies and programs with incentives for early adopters. Until that point, costs will drop through materials R&D and improving performance, not through premature volume production. The DOE will only encourage widespread adoption of FCVs when the technology meets customer needs and expectations, and when industry can establish a business case for large scale production and investment in hydrogen infrastructure.

Hydrogen fuel cells may become commercially competitive in some stationary, mobile, and auxiliary applications earlier than for transportation. The experience gained in these markets will help speed and encourage fuel cell research for vehicles. The DOE supports development of fuel cells for these applications, while focusing our activities and long term goals on the light duty transportation sector—among the most challenging, but also most important, energy sectors the DOE must address to meet our energy security and environmental goals.

Figure 9-2 summarizes the DOE short term hydrogen transition program from 2000 through 2015. The figure shows how the different learning, demonstration, and technology validation projects funded by the DOE will play different roles in different phases of development of the hydrogen fuel cell pathway.

Some members of the auto industry indicate they plan to produce a commercial fuel cell vehicle before 2015, perhaps as early as 2010. This is an admirable goal, and one which the DOE is working to achieve as well. However, the DOE's 2015 commercialization go/no go decision point is not based solely on the ability to build a vehicle. Hydrogen FCV prototypes exist today; the challenge lies in establishing them as feasible commercial vehicles for a mass audience. It may be possible to introduce FCVs to a small community of early adopters, or to controlled fleets with different needs than those of the general population. Today, controlled fleet demonstration projects allow FCV manufacturers to demonstrate their technology for the military and companies such as Federal Express and United Parcel Service. However, for widespread commercial acceptance, the full range of our technical and commercial targets must be attained, a goal that we anticipate reaching in 2015.

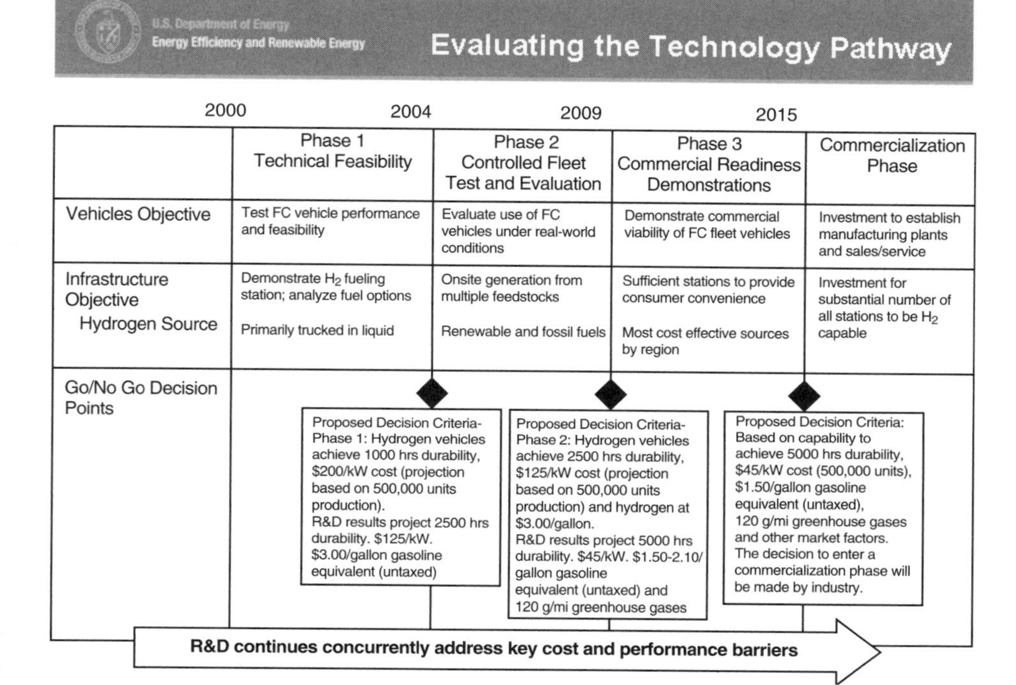

FIGURE 9-2. The DOE pathway to hydrogen through 2015.

The 2015 decision date also takes into account progress in overcoming economic and institutional barriers, including:

- The high investment risk of developing a hydrogen delivery infrastructure, given technology status and current demand
- The lack of uniform codes and standards to ensure safety, insurability, and fair global competition
- The low level of education regarding hydrogen safety and benefits among local code officials, policymakers, and the general public

The technology validation element of the DOE Hydrogen Fuel Cells and Infrastructure Program is shown in Fig. 9-3. It confirms that component technologies can be incorporated into a complete system solution, and that system performance and operation are met under anticipated operating scenarios and under realistic operating conditions. The results of the validations will be used to provide feedback on progress, to efficiently manage the research elements of the program, and to provide redirection as needed.

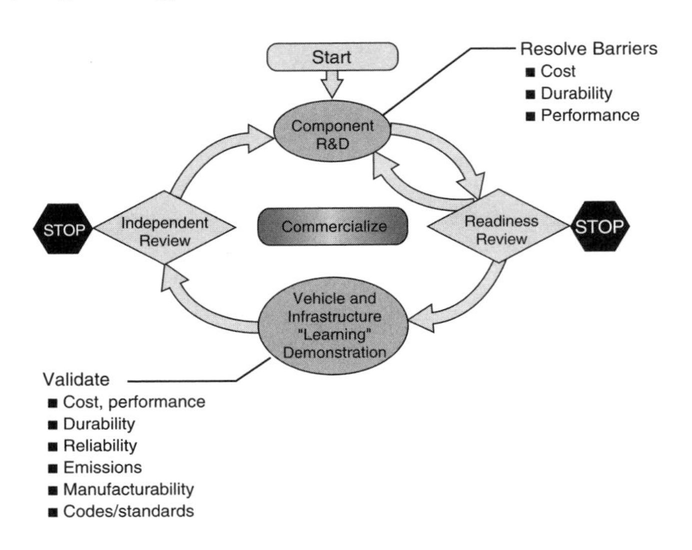

FIGURE 9-3. Technology validation cycle for hydrogen.

Why Is Hydrogen a Viable Alternative Fuel for the Future?

Hydrogen has the potential for both significant displacement of petroleum based fuels and zero or near-zero emission vehicle operation. While most current alternative fuels provide these same benefits, the potential benefits to both are greater with hydrogen. Historically, issues concerning fuel supply infrastructure, cost, and vehicle performance have slowed the growth of alternative fuels in the marketplace. The DOE has focused its efforts on addressing each of these issues through concerted research efforts with the fuel and automotive industries. These ongoing efforts will directly benefit hydrogen as a future viable alternative fuel. For example, the DOE's support of natural gas fuel infrastructure for both compressed and cryogenic liquid fuels will directly support the development of future hydrogen fuel supply infrastructure. Further, many of the domestically produced alternative fuels like natural gas and ethanol can be hydrogen carriers from which vehicle fuel grade hydrogen can be produced.

As pressure to reduce vehicle emissions has increased, industry has continued to seek solutions, including the use of alternative fuels. Partnering with federal agencies to develop alternative fuels was a logical and necessary step for industry to meet shared national and regional goals of a low-emission transportation system. The DOE recognizes the important role of supporting industry efforts through basic R&D in fuels and technologies that could achieve these air quality goals while at the same time reducing dependence on foreign energy. The DOE's previous and ongoing support of alternative fuel technologies has created a bridge for a viable and sustainable hydrogen economy.

While technical, institutional, and political issues remain to be addressed, the combination of previous DOE programmatic and research support of alternative fuel technology, the fuel and automotive industry commitment to commercialization, and projected broad U.S. market acceptance ensures hydrogen's place in the future transportation fuel market.

Industry Support for Hydrogen and Fuel Cell Technology

The automotive and energy companies, two very important industry sectors, have shown sustained interest in hydrogen and fuel cell technologies. Without such a commitment from the private sector, no program aimed at commercialization can be successful. This need for industry participation is why the DOE has worked closely with these industries to cooperate on research, development, and demonstration of key technologies.

Many of hydrogen's public benefits are also private benefits. Hydrogen, potentially a zero emissions fuel, is attractive because it frees automakers from many concerns about vehicle emissions and pollution. Repeatedly readjusting production to meet stricter standards can be expensive. Rather than continued redesign of vehicle product lines to meet mandated incremental emission reductions, the industry sees the use of hydrogen and fuel cells as eliminating this costly cycle from the vehicle production process. It is desirable to preempt regulation, and, in the words of Larry Burns, senior vice-president of General Motors, "remove the automobile from the environmental equation" (Burns, 2003). The industry views hydrogen and fuel cell technologies as viable long term solutions to the vehicle emissions problem.

In addition, unlike some alternative fuels which have limited applications as a result of feedstock supply, geographic, or other technical reasons, hydrogen FCVs have the potential to replace light duty gasoline ICEs with a vehicle powered by a widely available and sustainable fuel. It is more attractive for both the government and the private sector to invest in a technology with a large commercial audience, rather than one with a small niche market.

Another advantage is that hydrogen can be produced from a variety of feedstocks, including renewable sources, nuclear energy, natural gas, coal, and other fossil fuels. This diversity of feedstocks allows hydrogen to be produced with fewer concerns about price shocks, foreign disturbances, or other disruptions to supply. Hydrogen unifies the many energy resources in the United States into one fuel.

Unlike gasoline and some alternative fuel mixes whose formulation can differ from nation to nation, and even state to state, hydrogen has the potential to be a uniform fuel around the world. This is an advantage for automobile and energy companies, which increasingly must consider the demands of the global market. Hydrogen produced through electrolysis is a pure, uniform fuel, and hydrogen produced from fossil fuels could also be purified to meet international standards to ensure maximum performance of proton exchange membrane fuel cells and global compatibility. In addition, automakers can use hydrogen in vehicles with either standard ICEs or in FCVs.

Commercial appeal is arguably the single most important factor in the success of hydrogen FCVs. Many important business partners from the energy and automobile industries have already demonstrated a committed interest in hydrogen and fuel cells, and launched hydrogen research programs. Major energy companies have worked to redefine themselves as providers of energy services, unlimited to the sale of any particular commodity; as BP, or British Petroleum, says, "Beyond Petroleum." The energy companies have experience managing large financial resources and capital, skills important for a long term research program like the development of a global hydrogen infrastructure. ChevronTexaco cites their company's long history in developing large scale infrastructure as experience that will help them to succeed in the hydrogen economy as they have in the hydrocarbon economy.

The automobile industry has also shown interest in the business opportunities a hydrogen fuel cell economy may someday offer. Auto executives are optimistic about their abilities to manufacture FCVs that meet or exceed their customer requirements, with features like quiet and smooth electric drive systems, auxiliary power options, greater flexibility of design, and more convenient refueling options, such as home-based hydrogen appliances. FCV designs may also allow manufacturers to reduce the number of parts in a car, thereby lowering costs and possibly increasing durability. While still acknowledging that a number of tough technical barriers—hydrogen storage, for example—must be overcome, many car companies have shown considerable enthusiasm for hydrogen and fuel cell technology, and are voluntarily supporting initiatives to develop and demonstrate such technologies, in collaboration with the FreedomCAR program and the California Fuel Cell Partnership.

General Motors has publicly stated that they plan to be the first company to sell 1 million fuel cell cars, and say they hope to begin offering fuel cell vehicles in the next decade (Bak, 2003). Ford expects to have its vehicles on the road by 2020, and expects half of the vehicles they sell in 2040 to run on hydrogen (Durbin, 2003). The enthusiasm of the private sector will also help the DOE educate the U.S. public about hydrogen and fuel cells.

The International Partnership for a Hydrogen Economy

Energy security and the environment, the two main drivers for the hydrogen economy, are both global problems, calling for global solutions. The interest in hydrogen from around the world is yet another reason to be optimistic about its potential. Japan's fuel cell and hydrogen technology research, development, and demonstration program has tripled in size since 1995. The European Commission recently announced a program of €2.1 billion (euros) on hydrogen, fuel cells, and other related renewable energy activities.

Iceland has committed itself to becoming the world's first total hydrogen economy. Developing countries, including China and India, have also expressed interest in hydrogen, and have initiated research programs themselves (International Energy Agency, 2003).

Recognizing that international cooperation and collaboration will be important to efficiently achieve hydrogen and fuel cell technology goals, DOE Secretary Spencer Abraham, in mid-2003, called for an International Partnership for the Hydrogen Economy (IPHE) to establish cooperative R&D efforts and the sharing of information necessary to develop a hydrogen economy. The international partnership will promote common codes and standards that will avoid trade barriers and promote global harmonization, while minimizing duplication of R&D efforts. Because the United States is the world's largest economy, it is in an excellent position to lead the way to a global hydrogen economy and to encourage other nations to work with us in achieving this transition. Secretary Abraham hosted the first IPHE Ministerial meeting in Washington, DC, from November 19–21, 2003.

Conclusion

While energy efficiency and adoption of alternative fuels can and are being encouraged as interim measures to combat energy and environmental problems in the United States, they are only partial solutions. Together, they don't come close to completely solving U.S. dependency on foreign oil, which is expected to grow from roughly 50 percent in 2003 to 68 percent by 2025. This dependence, and the related challenges of air pollution and greenhouse gas emissions, will not go away. Energy independence has been a goal for decades, but without a truly viable means of replacing the tremendous quantities of oil imported to the United States, political will can only go so far. Ultimately, businesses and consumers will opt for the cheapest, most reliable fuel, which for decades has been petroleum. In the future, that fuel could be hydrogen.

As a first step, the R&D program at the DOE is needed to more firmly establish the technical and economic feasibility of hydrogen and fuel cells, specifically their ability to function with performance and cost comparable to conventional vehicles. Without this assurance, the private sector will be unwilling to invest in the hydrogen infrastructure necessary to make the hydrogen transition possible. The government can also help develop important codes and standards, and help build the human infrastructure necessary to ensure hydrogen safety. Additional policies beyond R&D can be implemented when a stronger political and scientific consensus exists to move forward.

With cooperation and enthusiasm from the private sector, the United States will develop technologies capable of eliminating our dependency on foreign oil, protecting the environment, and generating profits. With collaboration and enthusiasm from our international partners, the United States will reach our technology goals faster; we will foster a fair and open technology transfer and trading system, coordinate integrated codes and standards, and build a supportive policy framework.

Hydrogen is not a panacea for all the world's energy ills, but it is a promising technology. With the interest shown by businesses, scientists, educators, and policymakers from around the world, we believe it is a feasible solution to our energy security and environmental goals. By working together, it will be possible, in the words of President Bush, for "the first car driven by a child born today to be fueled by hydrogen, and pollution-free" (Bush, 2003).

References

Bak, Erik Poul. 2003. "Hybrids and Hydrogen Are the Stars of General Motors' Sacramento Tech Tour." Based on press release from GM. February 10, 2003. http://www.h2cars.biz/artman/publish/printer_92.shtml.

Burns, Larry (representing General Motors). 2003. Presentation at Asilomar Biennial Conference. Monterey, CA. August 1, 2003. See also on-line attributed to Rick Wagoner, CEO of General Motors at http://lang.motorway.com/home/articles/autonomy.asp.

Bush, George W. 2003. Presidential State of the Union Address. Washington, DC. January 28, 2003. http://www.whitehouse.gov/news/releases/2003/01/20030128-19.html.

Durbin, Dee-Ann. 2003. Ford Sees Future with Hydrogen. *Fuel Cells Today* (June 10, 2003): http://www.fuelcelltoday.com/FuelCellToday/IndustryInformation/IndustryInformationExternal/NewsDisplayArticle/0,1602,3007,00.html.

International Energy Agency. 2003. "Toward Hydrogen." IEA Workshop Report. March 3, 2003. http://www.iea.org/dbtw_wpd/textbase/work/workshopdetail.asp?id=98.

CHAPTER 10

The Hydrogen Transition: A California Perspective

James D. Boyd

When I was asked to prepare remarks for presentation at the 9th Biennial Asilomar Conference on Transportation and Energy, four questions were posed for me to address in the context of California's energy future. These questions were:

- What is the role of different levels of government (research, demonstrations, incentives, regulations, other policies) in the transition to hydrogen?
- When uncertainty dominates, what policies make sense?
- Are there models for successful policy intervention of this magnitude and duration?
- Timing issues—what to do now and what to do later?

After listening to the discussions at Asilomar, it seems these questions were a bit too presumptuous. Many experts question either the need for a hydrogen transition, or are concerned that an aggressive effort to promote this transition today is not right in the current political, economic, social, and technology era in which we are now living in the United States. With this in mind, I propose four alternative questions which might be more appropriate to define today's hydrogen debate. These new questions could be:

- Are we rushing headlong toward a hydrogen future?
- Do we know how to get there?
- What have we learned from past programs that could help us better understand the path to hydrogen?
- Are there better options? For example, are battery electric vehicles (BEVs) back in the race?

Having posed these alternative questions, let me provide a California context for a hydrogen transition, including what we have done in the past, what we are doing today, and what we still need to address. There are now more than 34 million people living in California, with 7 million more expected by the end of the decade. This huge population faces energy and environmental challenges that must be addressed even before a single commercial hydrogen fuel cell vehicle is likely to arrive in an auto showroom.

California, if taken as a country, would be the fifth largest economy in the world, the fifth largest consumer of energy, and the fifth most efficient energy consuming economy. Great strides have been made in improving energy efficiency in California. The state now has the lowest per capita electricity consumption among all 50 states in the United States. Progress has not been so impressive in the transportation sector, which ranks number 37th in energy use per capita, but number one in total gasoline and diesel fuel consumption. Transportation uses one-half of all energy consumed in California each year, including some 13.8 billion gallons of gasoline and 2.4 billion gallons of diesel.

How has California addressed the key hydrogen transition questions? Let's take them one at a time.

Are We Rushing Headlong Toward Hydrogen?

To answer this question for California, we have to ask what is broken that needs to be fixed quickly? Poor air quality, lack of energy security, and high greenhouse gas emissions are the three major issues influencing transportation energy policy in California today.

Poor air quality due to high motor vehicle emissions is perhaps the top environmental and health priority for Californians. Conventional transportation fuel use dominates this problem, accounting for roughly two-thirds of urban air pollution in many California cities. Although a 20-year effort to reduce automotive pollution has scored major successes in reducing pollution per vehicle in areas such as Los Angeles, urban growth, increased car ownership, and increased travel times have led to a worsening of air quality in many other cities, including those in the fast growing Central Valley.

As the results of a June 2003 survey by the Public Policy Institute of California (PPIC) suggest, most Californians know that motor vehicle exhausts are a threat to their health, and they are prepared to do something about it (Public Policy Institute of California, 2003). Fifty-eight percent of Californians surveyed believe that air pollution is a serious health threat to them and their immediate family, and 37 percent say they or a family member suffers from asthma or other respiratory problems. Accordingly, 30 percent of state residents rate air pollution as the most important environmental issue, followed distantly by water pollution with 10 percent, or growth and sprawl with 7 percent.

Sixty-five percent of Californians surveyed by PPIC say they would be willing to support tougher air pollution standards on new cars, trucks, and sport utility vehicles (SUVs), even if it raises the cost of buying a vehicle.

Three-quarters favor requiring automakers to significantly improve the fuel efficiency of cars sold in the United States, even if it increases the cost of a new car. A large majority—79 percent, including 69 percent of SUV owners—also say they favor changing federal regulations on SUVs to match existing fuel economy standards for passenger cars. Eighty-one percent of Californians support giving tax breaks to encourage consumers to purchase advanced electric vehicles. How to mobilize this knowledge and enthusiasm for change in support of a hydrogen transition is a big challenge to hydrogen advocates today.

A historical sidebar about the California Low Emission Vehicle Program is appropriate here. The program dates back to 1986 when the California Air Resources Board (ARB) staff discussed and debated the fact that we could not achieve the 1987 Federal Clean Air Act public health protection target in spite of doing more than anyone anywhere to achieve clean air. In spite of the auto industry's assertions that they had already been driven to near-zero, more had to be done to reduce vehicle emissions. In fact, the ARB felt that California would never see clean air unless some percent of the vehicle population's emissions were, in effect, zero. But zero had to wait for the unveiling of GM's Impact electric vehicle prototype, the precursor to the EV-1, in 1990. This need, zero, is still true today.

Concern over rising greenhouse gas emissions is a second policy driver in California. Again, transportation fuel use dominates this problem in California. In 2003, the transportation sector contributed 58 percent of total California carbon dioxide (CO_2) emissions, which accounted for 1.5 percent of the world's total greenhouse gas emissions.

In July 2002, the California legislature enacted Assembly Bill (AB) 1493, a landmark act that places the state at the forefront of a worldwide effort to reduce transportation generated greenhouse gases. AB 1493 is the first law in America to substantively address the threat of global warming. The law requires the ARB to develop greenhouse gas emissions standards for vehicles in model year 2009 and beyond. The standards will apply to automakers' fleet averages, rather than each individual vehicle. Californians will continue to choose and purchase vehicles of their choice. Although California's emphasis on reducing greenhouse gas emissions from transportation is not shared by the policies of the current U.S. national government, it is widely supported and mirrored by the policies of nations around the world.

Finally, energy security concerns, highlighted by the frequent episodes of high gasoline and diesel price volatility, have become a pressing problem for Californians and the state's economy. Demand for petroleum fuels is growing at nearly 2 percent annually, and California no longer can meet its gasoline and diesel fuel demands with in-state refining capacity. The trend toward importing more crude oil in general and finished oil products specifically is disturbing. Finished fuel products or blending components are even more subject to the vagaries of world events than crude oil.

Moreover, California's just-in-time motor vehicle fuel supply system is disrupted regularly because of upsets and breakdowns that occur at the few refineries that still operate in the state. The result is a fuel supply system that is as taut as a stretched rubber band.

To ease oil supply problems, the California Energy Commission (CEC) commissioned studies on the feasibility and advisability of a state sponsored gasoline product pipeline from refineries along the Gulf of Mexico, and establishing a state sponsored strategic fuel reserve. While advising against both solutions, the CEC identified a number of problems possibly impairing the state's ability to import fuel and blending components and will work with all parties to rectify these issues.

Another new study, "Reducing California's Petroleum Dependence," by the CEC and the ARB (2002), addresses ways to reduce California's dependence on petroleum. It was requested by the state legislature, and has been completed and approved by the governing boards of these agencies and submitted to the governor and legislature. The legislative impetus for this study was the volatility of fuel prices being faced by Californians. The report analyzes fuel efficiency, fuel substitution, pricing measures, and other modal options. The report recommends three California goals:

- Reduce California petroleum consumption to 15 percent below 2003 levels by 2020
- Double the fuel efficiency of new cars, light trucks, and SUVs
- Increase the use of nonpetroleum fuels to 20 percent of on-road fuel consumption by 2020 and to 30 percent by 2030

Our analysis shows that by far the most effective strategy to reduce petroleum consumption is increased new vehicle fuel efficiency. A secondary strategy involves the substitution of hydrogen fuel cell vehicles and other alternative fuel or advanced vehicles. These two strategies implemented together could yield significant motor vehicle emission and fuel use reductions in the 2020 to 2030 time frame.

However, we continue to ask ourselves questions. First, as a CEC commissioner, I constantly ask if a proposed action will forego options that are cheaper and as effective in reducing petroleum consumption. Secondly, I ask if the state is losing sight of near term issues regarding petroleum supply and demand in favor of long term possibilities that may or may not come to fruition. Finally, looking ahead to the hydrogen end game, am I taking the wrong approach in the transition to hydrogen?

The United States, contrary to California, has never really had an energy policy; it continues to sustain petroleum as the primary transportation fuel without effective market intervention. Except for the few "energy crises" that spurred interest in energy policy needs and created a few programs—crises that so far have been quickly forgotten—air quality concerns have carried the banner for fuel policy program needs.

Do We Know How To Get To and Through a Hydrogen Transition?

Many energy analysts, including myself, believe different levels of government policies and programs must play a vital role in a transition to hydrogen. It is the role of government to provide public goods. In accordance with a number of California statutes, government's role in the hydrogen transition, regardless of the level, should be to advance science and technology, and disseminate information of value to California not adequately provided for by the competitive and regulated markets. An important role of government is to continually assess, in a transparent and fair manner, all options available to our society to meet its goals.

The CEC has shouldered its responsibility, over time, by investing resources in transportation research, demonstrations, and incentive programs. Even during these times of limited resources, the CEC has continued to invest in infrastructure and application projects. It has sought crossover opportunities with stationary applications, e.g., distributed electricity generation. It is participating financially in a number of motor vehicle fueling infrastructure projects, including those at AC Transit, Valley Transit, Chula Vista, and Los Angeles International Airport. The CEC has a hydrogen infrastructure study underway exploring safety issues and steps to develop hydrogen stations and convert existing natural gas stations to provide hydrogen.

Other government agencies at the state and local levels within California, e.g., the ARB, local air districts, transit agencies, and the university system, are doing more. These are important and necessary activities. Research is a basic foundation for further action and activities.

The federal government also has major responsibilities to fulfill in this arena, encompassing the entire range of possible activities. The U.S. president's declaration in his January 2003 State of the Union address that we need to find an alternative to petroleum alone has saved untold dollars and hours educating the citizens of this country to the fact that we have to take action. The federal FreedomCAR program and fueling infrastructure initiatives are other more tangible contributions to the hydrogen transition.

Finally, never underestimate or sell short the role and value of government regulation. The deliberate but judicious use of government's regulatory authority can facilitate progress, provide for new technology, and otherwise provide cost effective, health protective, and socially beneficial products.

Multiple pathways are desirable. In my opinion, particularly in the area of energy, a portfolio approach has proven itself both in the past and in the present. Hydrogen and hydrogen fuel cell vehicles face considerable challenges as they move into the marketplace. These include challenges regarding price, product performance, codes and standards, infrastructure costs, and public education, just to name a few. The collective size of these challenges suggests a constant look for synergies and areas of cooperation and partnership in the design and implementation of hydrogen policies and programs.

What Have We Learned from Past Programs?

Are there models for successful policy intervention of this magnitude and duration? In California, there are many! The path we have adopted in California includes a menu of actions to be taken in order to travel the road to a hydrogen future. These actions include vehicle efficiency improvements, alternatives to conventional fuels, greater penetration into the vehicle fleet of hybrid electric vehicles (HEV), and eventually hydrogen fuel cell vehicles in significant numbers.

The ARB's Low Emission Vehicle Program, including the requirement for zero emission vehicles, is classic. It has spurred scores of important technological advances, and has made major contributions to cleaning up California's air quality. Moreover, the CEC and ARB's pursuit of clean alternative fuels, methanol and natural gas in particular, has inspired the oil industry to bring to market the world's cleanest burning gasoline and diesel fuels. The Clean Diesel and Gasoline Fuels programs, working synergistically with vehicle controls, have yielded significant air quality and public health improvements.

The "Reducing California's Petroleum Dependence" report mentioned earlier gave the CEC and ARB a framework to examine various options and will undoubtedly foster debate on a number of policy questions affecting our transportation fuel future. The CEC's newly resurrected and continuous responsibility to deal with California's energy future through the biennial "Integrated Energy Policy Report" provides a continuous forum for the subject of the hydrogen transition (CEC, 2003).

Public–private partnerships are another pathway being pursued in California. The California Fuel Cell Partnership has facilitated precedent-setting collaboration between government and the energy and vehicle industries. CALSTART, born out of the sharp reduction in California's aerospace and defense industry and the concurrent existence of the Zero Emission Vehicle mandate, has given birth to new technologies and new companies in the transportation and energy fields. And let's not forget the California government, university, and industry collaborations that bring us the likes of the Institute of Transportation Studies at the University of California, Davis, and the National Fuel Cell Research Center at the University of California, Irvine.

Another lesson learned from the past decade is that regulatory programs do not always deliver what was expected. For example, methanol vehicle production did not boom as a result of a number of government programs designed to promote its use in California, but the mere threat of methanol as the dominant motor fuel brought cleaner burning gasoline to California. In contrast, no air quality benefits are resulting from a federal program that provides automakers with CAFE credits for selling flexible fuel vehicles capable of burning ethanol, but which do not do so because there is no ethanol fueling infrastructure and the cost of ethanol is too high.

Let us not forget to continually question what we have done and what we want to accomplish. Learn from history. For example, avoid picking winners prematurely. Government, in particular, should be continuously analyzing and assessing options. Other lessons are to facilitate research and development activities today—in this case thoroughly understanding the objectives of a long-term transition to hydrogen and the timeframes required to complete each critical step.

Are There Better Options Than a Transition to Hydrogen?

As compelling as the arguments for a rapid transition to hydrogen are, the energy and environmental problems in our transportation sector are enormously complex. General Motors learned a lot from its experience with the battery powered EV-1, the vehicle that proved a ground-up, attractive, and beautifully performing ZEV was possible, as did Aerovironment's Paul MacCready, the "father" of the Impact.

Paul MacCready argues forcefully in Chapter 16 in this book that prematurely picking hydrogen fuel cells could have major consequences in dampening research and development of advanced battery technologies, which he believes will soon be able to outperform hydrogen fuel cells, with fewer requirements for new fueling infrastructure.

Similarly, John DeCicco, from Environmental Defense, laments in Chapter 15 the lack of public policies to promote fuel efficiency. He believes fuel efficiency achieves most of the environmental objectives of the United States sooner and at a much lower total cost than commercializing hydrogen technologies. He also questions if the level of public support for a major transportation energy transformation is strong enough to sustain the effort for the many years it will take before hydrogen fuel cell vehicles begin to clean the air and reduce oil imports to the United States.

These speakers, and surprisingly large numbers of other energy experts, suggest to me that it is important for government leaders to resist picking winners or losers for the future U.S. transportation market.

Conclusions

Returning to hydrogen and policy options and governments' role, there is widespread belief that consumers rely on government to tell them there's a problem. A corollary to this theory is that consumers, the public, expect their government to do something about the problem. In California, the air quality and energy program efforts have been relatively successful. Government leadership, government research, public support, and positive response have been key. California's years of success in battling air quality problems is paramount. Big research investments, public education, and partnering with stakeholders all play major roles in California's success.

In closing, let me compare today's move toward hydrogen to other moments in history, and ask what is different this time. This time, the biggest energy and automotive companies are at the table with government agencies, infrastructure companies, and other stakeholders, talking about hydrogen, fuel cell vehicles, and all the programmatic needs that have to be addressed to complete the road to a hydrogen economy. At no other time have the energy and auto companies partnered to facilitate a fundamental change to an alternative fuel powered advanced vehicle technology. That's what is different, and that's what will make the difference in the transition to hydrogen.

References

California Energy Commission (CEC) and California Air Resources Board. 2002. "Reducing California's Petroleum Dependence." Report posted on CEC Web site: http://www.energy.ca.gov/fuels/petroleum_dependence/index.html.

California Energy Commission (CEC). 2003. "2003 Integrated Energy Public Policy Report." Report posted on CEC Web site: http://www.energy.ca.gov/energy policy/index.html.

Public Policy Institute of California (PPIC). 2003. Results of public opinion survey posted on the Web site: http://www.ppic.org.

CHAPTER 11

U.S. Hydrogen Activities—A European Perspective

Barend van Engelenburg

In my daily work, I help to create policies aimed at achieving a cleaner transport system in The Netherlands and in Europe. In April 2003, I arrived at Stanford University in California to study the experience in the United States in the transportation field, including the pursuit of hydrogen vehicles, at the federal and California levels. The goal was to return with practical lessons for The Netherlands.

While in the United States, I attended the 9th Biennial Asilomar Conference on Transportation and Energy, which helped shape many of my viewpoints acquired during my extended visit. In this chapter, I compare the U.S. approach toward hydrogen with the approach now being taken within the European Union (EU), preceded by a description of the European context.

A Short History of the EU Commitment

The European Union hydrogen commitment is similar to the United States in that it represents an economic and political collaboration of states. The differences, however, are much more pronounced. Individual European states are much more independent than states in the United States, and the parliament has much less formal power than the U.S. Congress. The executive branch of the European Union is called the European Commission (EC). The Commission is the only body with the right to propose EU-wide legislation through its "right of initiative." The European Council is the main decision-making body of the EU. The Council consists of national representatives at ministerial level. Formal laws in the United States are called directives in the European Union, but they do not have the same status as laws in the United States. A directive is formal text that will only become effective after implementation by member countries. Each country is free to enact the provisions of the directive in its own way according to national rules and habits. The Commission can enforce implementation, however. For more details see the references at the end of this chapter.

The European Union has a common and coordinated research and development program called a Framework Program (FP). At the end of 2002, the fifth program ended and the sixth program started. In FP5 the Commission contributed about €144 million (euros) to fuel cell or hydrogen projects (see Fig. 11-1).

In the FP process, the Commission identifies, after consultation with the member countries, the themes for research and development for a given period, usually 4 years. The Commission also defines a preliminary budget allocation for each theme. It then issues a call for proposals in the *Official Journal of the European Communities.* Each research consortium submitting a proposal must include research institutes from the European Union, but it may include non-EU research institutes or other partners. The proposals are evaluated and ranked, and the best-scoring proposals within the budget limit are rewarded with a contribution. That contribution for the most part does not exceed 50 percent of the total research budget. The research partners generally have to find more than 50 percent of the budget from other means, such as state government or industry contributions. Framework Program R&D programs in general represent only about 5–10 percent of the total research budget of the member countries.

Hydrogen first gained attention in the European Union in 2000 with the "Green Paper" on security of energy supply, when hydrogen was nominated

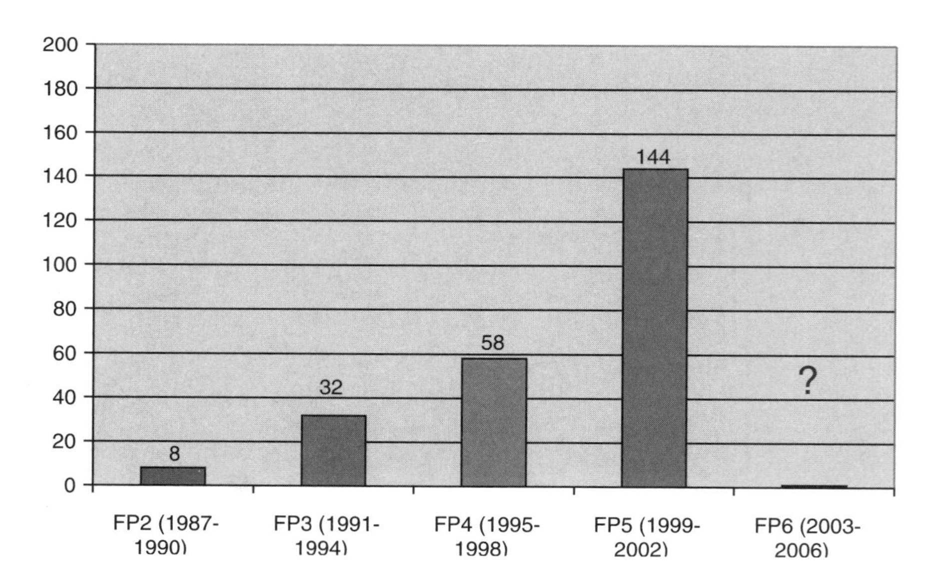

FIGURE 11-1. Commission support for fuel cell–hydrogen projects in Framework Programmes (in million €). FP6 data (*) apply to the first call only, see EC 2004.

as a possibly important fuel for transportation (European Commission, 2000). In the "White Paper" on transport policy (European Commission, 2001a), the Commission emphasized its willingness to deal with alternative fuels, including hydrogen. Two new directives on renewable fuels, including references to alternative transportation fuels, followed these papers at the end of 2001. Commission green papers are documents intended to stimulate debate and launch a process of consultation at the European level on a particular topic. Commission white papers are documents containing proposals for action in a specific area. In some cases they follow a green paper published to launch a consultation process at the European level. When the Council has favorably received a white paper, it can become the action program for the European Union in the area concerned.

The negotiations that followed release of the green and white papers revealed that most member countries were not yet ready for a practical step forward. Major disagreement remained about the goals of the directive on renewable fuels for transport. At the same time, the communication stimulated much debate about a proper approach for encouraging hydrogen. In the course of 2002, the Commission appointed two groups: the Alternative Fuels Contact Group (AFCG) and the High Level Group on Hydrogen and Fuel Cells Technologies (HLG). Both groups included high-level representatives from industry, research institutes, and end-user organizations. Both groups reported to the Commission in mid-2003: the HLG with its final report (High-Level Group for Hydrogen and Fuel Cells, 2003a), the AFCG with an interim report (Alternative Fuels Contact Group, 2003). In June 2003, an EU conference on hydrogen was held in which the Commission officials expressed firm commitments to hydrogen and signed a collaborative agreement between the Commission and the U.S. Secretary of Energy. In the summer of 2003, the Commission decided to fund more than a dozen projects on hydrogen in the, first call of FP6. One, the Hyways project, aims to deliver a European roadmap for hydrogen, both in transportation and stationary applications.

Comparing U.S. and E.U. Approaches

This comparison of the U.S. and EU approaches is divided into two issues: process and content. Comparing process means asking whether they have organized the process of choosing priorities, consulting stakeholders, and implementing decisions in the same or a different way. Comparing content means asking whether they focus on the same or different subjects and whether they chose similar or different short term activities or routes for implementation.

When looked at from a distance, the two processes appear similar. A broad consulting process with budget decisions as the final result appears in both. But looking at them more closely reveals large differences.

In the first place, in the United States, the Bush Administration is acting as the director of the process, while in the European Union the

Commission is acting more like a facilitator. The Bush Administration has apparently selected hydrogen as the long-term energy choice, and it is eager to direct the whole process. The Commission stated a much weaker commitment to hydrogen (e.g., European Commission, 2001b), reflecting differences in opinion about the hydrogen issue among the EU member countries.

Secondly, the United States used an open process consulting with diverse groups in conducting the vision and roadmap process. The EU High Level Group was closed and less diverse: no environmental organizations were included, for instance.

Finally, the United States has chosen a structured approach proceeding stepwise from a basic resolution to a vision, roadmap, implementation plan, and external review. The structure of the process in the European Union is not yet clear. There is no formal resolution or decision about hydrogen. One could describe the High-Level Group report as a vision document, but it is not presented as an official communication of the Commission to the EU Council. There will be a roadmap project, but that is an external research project that will not be structured and directed by the Commission. The EU processes might be characterized as having moved to a point equivalent to the step in the U.S. process somewhere between the recommendations of the National Energy Policy Development Group (NEPP) and the delivery of the vision document; the High-Level Group for Hydrogen and Fuel Cells (2003a) can be considered the beginning of a vision document.

In short, the U.S. process can be characterized as a strongly directed, relatively open process with many stakeholders involved and with a predefined and fairly logical structure. The EU process might be characterized as ad hoc and fragmented. It is inspired more by efforts in the United States and Japan, and by fears of lagging behind, than by a clear vision of a hydrogen economy.

Turning to a comparison of content, I focus on fundamentals, focus, and practical activities. Under the heading of fundamentals, I compare basic attitudes toward hydrogen. For example, what kind of end vision for hydrogen production is given and why should hydrogen be pursued in the first place?

The difference between the European Union and United States in fundamentals or attitudes is not clear. Both seem to think that hydrogen can be a solution to many different problems. But why is hydrogen a better solution for those problems than other alternatives? Figure 11-2 shows a list of possible alternatives for the long term, all of them able to meet most of the relevant sustainability criteria. One could, therefore, expect an answer to the question: Why hydrogen and not a portfolio approach?

The European Union commissioned an assessment of alternatives that focused attention on fuel cells and hydrogen, which included the following statement:

> *Rationalising the use of conventional private cars in town and city centres and promoting clean urban transport are also priority objectives and likewise efforts towards using hydrogen as the fuel for vehicles of the*

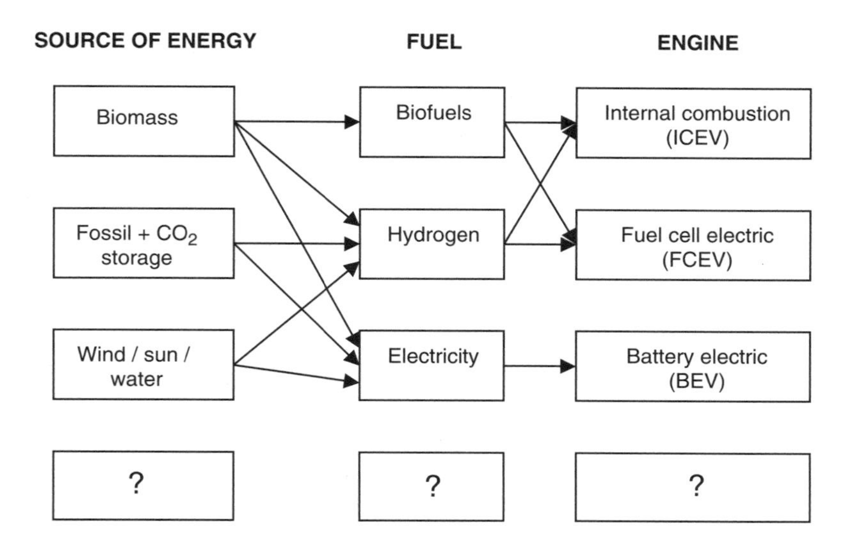

FIGURE 11-2. Fuel chains with long-run potential for sustainable mobility. *Source:* Ministry of Housing, Spatial Planning and the Environment (2003).

> *future. Among the initiatives might be the promotion of the commercialisation of zero or low polluting vehicles (for private and commercial use). The development of a new generation of electric, hybrid (electric motor combined with thermal motor) or gas powered cars or, in the long term, vehicles running on fuel cells are also very promising.*
>
> *European Commission, 2000, 71*

But the analysis is loose, and not complete nor convincing. In U.S. documents, there also does not seem to be a careful assessment of alternatives. This is no surprise since most of the U.S. documents are based on a political determination by the Bush Administration, with the NEPP as the basis for the decision. No comparative assessment, at least in the text of the NEPP, was made. In short, the answer to the basic fundamental question in both regions is not very convincing.

There also seems to be agreement about a certain part of the end vision of the hydrogen economy. The hydrogen recommendations of the NEPP are given in the chapter "Nature's Power: Increasing America's Use of Renewable and Alternative Energy." This seems to imply a certain connection of hydrogen to renewable resources. In the relevant Commission documents, as well as in the report of the HLG, there is also a connection to renewable resources. See, for example, the following quotes:

> *In the longer term, the possibilities for other renewable sources of fuels, such as hydrogen, need to be exploited. ... Research in this sector should*

> *be intensified, with a view, notably, to explore new solutions linked to the utilisation of alternative energy sources, such as hydrogen (which, together with methanol, is the fuel used in fuel cells and which can be produced from several sources of primary energy, including renewable sources).*
>
> *European Commission, 2000, 44*

> *Sustained efforts should be made to promote the penetration of new and renewable energy sources (such as hydrogen and co-generation) in our economies. The European Union has set itself an ambitious target in this respect: 12% of energy consumption in 2010 should come from renewables.*
>
> *European Commission, 2000, 72*

This agreement is, however, only superficial: In reality the U.S. documents mostly characterize hydrogen as an alternative fuel. There is not a close link between hydrogen and renewable resources in the NEPP. In fact, the possibility of producing hydrogen from a wide variety of sources, including fossil and nuclear energy resources, is one of two main attributes of hydrogen that motivated U.S. interest in hydrogen (Garman, 2003). The political emphasis in the European Union is indisputably renewable sources. Many members of the European Parliament have emphasized this focus; it is also advocated by Fuel Cell Europe, a lobby group for fuel cell research and development. The High-Level Group (2003a), however, gives a more neutral and diverse perspective, more like that of the Department of Energy (2002a).

It is evident that a careful assessment has not been made of lessons already learned from previous alternative fuel policies. Chapter 12 in this volume by McNutt and Rodgers discusses the consequences in the United States of earlier policies that emphasized using certain niche markets. They suggest that current hydrogen programs would have been designed differently if those lessons had been taken into account. The situation in the E.U. is the same: hardly any connection to lessons learned in past policies.

The starting points of the hydrogen energy plans in the European Union and the United States are similar. The challenge is to define steps that take the uncertainties into account and stimulate early introduction. This is not an easy task. Part of the focus process is to identify the technical barriers: What are the main technical bottlenecks that prevent these technologies from being commercialized? Such an analysis is aimed at achieving a research and development agenda. Comparing the European Union and the United States leads to the conclusion that very high priority needs to be given to cost reduction of fuel cells and development of hydrogen storage technologies, and greater attention to production and infrastructure issues and application experiments. There is some difference in emphasis. In the United States, coal combined with carbon dioxide sequestration and nuclear energy are part of the portfolio right from the start. The European Union focuses more on renewables, although the HLG (2003a) mentions coal and nuclear options.

In both regions, the vision and roadmap documents stress an integral and connected approach for research and development, but are less clear about market development. The European Union did not address market issues early on, although a May 2003 EU directive does promote alternative fuels for transport (European Commission, 2003). That directive is planned to be a stepping-stone aimed to stimulate a market pull for alternative fuels. The U.S. implementation programs address market issues, but actual policies and programs have not been proposed. My preliminary characterization of the U.S. implementation plans would be high priority for research and development and low priority for market development.

Until now, most activities in the United States and the European Union have been related to research and development. In the United States, those activities are mainly directed by the DOE, and in the European Union by a consortia of collaborating industries and academia. In the last few years, there has been more attention in both regions to vehicle and fuel demonstrations. The European Union launched a very visible project called Clean Urban Transport for Europe (CUTE). The Commission contributes €18.5 million to this project costing a total of €52 million. In this project, 30 hydrogen buses will drive in 10 large European cities. This project is designed as an experiment aimed at learning by practical experience. At the same time the Commission also contributes to a hydrogen project in Iceland, called Ecological City Transport System (ECTOS). This project seeks to learn more about the practical issues related to hydrogen buses, including production, safety, and commercial application. Until recently, in the United States most of the practical projects have been carried out in California, as discussed by Boyd in Chapter 10 in this volume. But Chalk and Inouye show in this book (Chapter 9) that large demonstration projects are on the drawing board as part of FreedomCAR, FutureGen, and the Hydrogen Fuel Initiative. By 2006, there will probably be a large portfolio of practical projects in both regions. It is hard to predict whether there will be any significant difference between the EU and U.S. portfolios.

Conclusions

Every innovation has at least two challenges: the technical challenge and the market challenge. The technical challenge is about getting the technology working and fit for consumer specifications. The market challenge is to get the innovation deployed.

The technical challenge for hydrogen exisits in every domain from production towards end-use and storage. The challenges expressed by DOE (see chapter 9 by Chalk) do not lack ambition. But, what is the chance that these challenges will be successfully met? What is also not fully understood is the benefits of hydrogen technologies compared to their competing alternatives, like the hybrid electric propulsion with internal combustion engines, advanced bio-fuels from lignocellulosic biomass and so on. The experts at least differ in opinion: DOE (2002b) can be placed on the optimistic side, Keith (2003) is far less optimistic and Pelkmans (2003a and 2003b) and the National Research Council (2004) are in between.

The market challenge is trying to answer the questions how key players can reach wide scale hydrogen deployment starting from their situation. The hydrogen case dose not present an easy case. The main problem is a chicken and egg dilemma; real deployment of a hydrogen conversion system like a fuel cell needs a wide and diffuse spread of hydrogen supply and the development of a supply system needs large demand for hydrogen. In the hydrogen case, this is not a simple dilemma that can be solved by a single intervention of government. Ulutas (2003) analyzes the diffusion of fuel cells in the U.S. based on a multilevel model and comes to the conclusion that large scale introduction of fuel cells and hydrogen is not an incremental change to the system and, therefore, there is a mix of opportunities and barriers but "as yet there are more barriers than opportunities." (p. 122)

The challenge then is to define the first steps that take into account the uncertainties and also lead to stimulating the introduction. This is not an easy task. It is not clear if government policy should start with research and development (R&D) or with the first, small steps towards implementation.

When examining the world today and its need for a transition to hydrogen, it appears that achieving the hydrogen economy will need a major system change with unclear costs and benefits. Thus, the present situation is not a very good starting point at all. In fact the DOE (2002b) also sketches some of these conclusions, although it does not elaborate on the uncertainties. For other comparable conclusions see HLG (2003a), Pehnt (2003) and Pelkmans (2003a and 2003b).

Even so, the main governmental actors in the U.S. and in the EU have chosen to enter the challenge. I conclude from my comparison of the EU and U.S. hydrogen programs that each region has its own approach and priorities. In the case of the management of the process and the relation of hydrogen to renewable resources, the differences are very large. The differences are significant enough to make collaboration between the European Union and the United States more than a trivial undertaking.

Most of my interactions with U.S. experts during my travels through the United States consisted of discussing the choice for hydrogen. I have heard visions of the hydrogen future range from extremely devout and zealous to radically cynical. If you add large media coverage combined with uncontrollable expectations and large uncertainties to that mixture of visions, you do not have a recipe for a solid foundation of a hydrogen policy.

I close with the question: Is a more modest governmental commitment to hydrogen either in the European Union or United States preferred as a way to face the challenges ahead? If my understanding of the U.S. situation is correct, then the U.S. answer is yes, possibly. According to our analysis in The Netherlands, hydrogen should be included in whatever portfolio of solutions emerges, not because it is the ultimate answer, but because we cannot afford to neglect any clean opportunity. Accepting this analysis leads to the following conclusions:

- One should be more modest and honest about the attributes of and goals for hydrogen
- One should focus more on solving problems than on promoting solutions, and thus be more relaxed in distributing experiments around the world
- One should focus on analyzing and understanding comparative advantages of different regions
- One should accept uncertainty and learning as the foundation, and use lessons from the past to promote future innovation

This recommended approach would be a more solid basis for international collaboration than shared technology development, with its inherent problems of intellectual property and conflicts of interest.

I came to the United States to learn from its experience so that we can improve strategy in The Netherlands. I have learned that solving global problems requires a realistic approach focused on solving that problem, not on promoting the solution with the highest sex appeal. I hope the United States recognizes the mirror image.

References

Alternative Fuels Contact Group. 2003. *Interim Report*. Brussels, Belgium: European Commission, March 31, 2003.

European Commission. 2000. "Towards a European Strategy for the Security of Energy Supply." Green Paper, adopted by the EC on November 29, 2000, COM (2000) 769 final.

European Commission. 2001a. "European Transport Policy for 2010: Time to Decide." White Paper, adopted by the EC on September 12, 2001, COM (2001) 370 final.

European Commission. 2001b. "Communication on alternative fuels for road transportation and on a set of measures to promote the use of biofuels." COM (2001) 547.

Garman, David K. 2003. Testimony of Assistant Secretary Energy Efficiency and Renewable Energy before the Committee on Energy and Natural Resources of the U.S. Senate. "Energy Use in the Transportation Sector." March 6.

High-Level Group for Hydrogen and Fuel Cells. 2003a."Hydrogen Energy and Fuel Cells: A Vision for Our Future." EC report, June, 2003.

High-Level Group for Hydrogen and Fuel Cells. 2003b. Conference on "The Hydrogen Economy—A Bridge to Sustainable Energy" organized around the presentation of the High-Level Group report "Hydrogen Energy and Fuel Cells: A Vision for Our Future." Brussels, June 16–17, 2003b. Proceedings at: www.cordis.lu/fp6/sustdev_h2.htm.

Keith, D. W., and A. E. Farrell. 2003. "Rethinking Hydrogen Cars." *Science* 301 (July 18): 315–316.

Ministry of Housing, Spatial Planning and the Environment (abbreviated as VROM). 2003. *The VROM vision on sustainable development*. In Dutch. Den Haag.

National Research Coucil, 2004. *The Hydrogen Economy: Opportunities, Costs, Barriers, and R&D Needs.* Committee on Alternatives and Strategies for Future Hydrogen Production and Use, National Academies Press.

Pehnt, M., and S. Ramsohl. 2003. "Fuel Cells for Distributed Power: Benefits, Barriers and Perspectives." Commissioned by World Wildlife Fund, in co-operation with Fuel Cell Europe, June 2003.

Pelkmans, L., S. Hultén, R. Cowan, G. Azkárate, and P. Christidis. 2003a. "Trends in Vehicle and Fuel Technologies—Overview of Current Research Activities." EC Joint Research Centre, IPTS, EUR 20747 EN (May 2003).

Pelkmans, L., S. Hultén, R. Cowan, G. Azkárate, and P. Christidis. 2003b. "Trends in Vehicle and Fuel Technologies—Review of Past Trends." EC Joint Research Centre, IPTS, EUR 20746 EN (May 2003).

Ulutas, Ö. 2003. "Diffusion of Stationary Fuel Cells in the US." In Dutch. Master's thesis. Technical University Eindhoven, September 2003.

Resources

Links for European Union and Commission information

General information:
http://europe.eu.int/abc/index_en.htm
Research:
http://europe.eu.int/pol/rd/index_en.htm
http://europe.eu.int/scadplus/leg/en/s23000.htm
FP6:
http://www.cordis.lu/fp6/
http://europa.eu.int/comm/research/fp6/index_en.html
HLG vision report:
http://europa.eu.int/comm/research/energy/pdf/hlg_summary_vision_ report_en.pdf

CHAPTER 12

Lessons Learned from 15 Years of Alternative Fuels Experience—1988 to 2003

Barry McNutt and David Rodgers

United States President G. W. Bush unveiled his administration's Hydrogen Initiative in 2003 with the following: "Tonight I am proposing $1.2 billion in research funding so that America can lead the world in developing clean, hydrogen-powered automobiles." Since then, private sector and public sector interest in hydrogen fuels has increased dramatically and the federal program has grown to $1.7 billion.

As U.S. Department of Energy (DOE) policy analysts during the years before the hydrogen initiative was launched, we have examined what the United States has learned from its long experience with alternative fuels. This chapter explores barriers that faced developers of these other alternative fuels and technologies that may also face developers of hydrogen and hydrogen fueled vehicles. It reviews the successes and failures of technology and policy options that have already been used to promote a transition to new fuels and vehicles. We hope this chapter may be a valuable resource for all those planning and investing in the hydrogen future.

Why Alternative Fuel Experience Is Valuable

From one perspective, hydrogen is just another in a long line of alternative transportation fuels. The U.S. Congress included hydrogen in the list of alternative fuels that qualified for special support and promotion through fleet programs as far back as the Energy Policy Act (EPACT) of 1992.

Hydrogen will probably face many of the same obstacles, if not more, as other alternative fuels. Some argue it won't. Some describe hydrogen as the "perfect" fuel because it contains no petroleum and emits no pollutants when consumed in fuel cell vehicles. Others see an historical, almost evolutionary, reduction in the use of carbon in fuels, leading inexorably to hydrogen as the "ultimate" fuel (Cannon, 1995). The Department of Energy has identified hydrogen as one of the keys to a secure, clean, and prosperous energy sector that will continue for generations to come (U.S. Department of Energy, 2003).

Our goal is not to refute these hydrogen proponents—they could be correct. Yet it seems apparent to us that the use of hydrogen as a transportation fuel faces many, if not all, of the same obstacles faced by other nonpetroleum fuels. These obstacles include the following:

- Lack of refueling infrastructure
- High cost
- Lack of vehicles engineered to operate on the fuel
- Difficulty breaking into an established market
- Perceived or real issues of safety and reliability
- Lack of driving range

In the case of hydrogen, some of these obstacles are even more of a barrier than for other alternative fuels. Hydrogen fuel, when used in its compressed form, shares many similarities with compressed natural gas (CNG). When liquefied, it shares many similarities with liquefied natural gas (LNG) and liquefied petroleum gas (LPG), also called propane. In addition, hydrogen is costlier to compress and harder to store than any of these other alternative fuels. Hydrogen fuel vehicles may, in their early years, have low range, a negative attribute that currently limits use of battery electric vehicles (BEVs).

Hydrogen fuel vehicles also must surmount a very difficult marketing challenge because there are currently no commercial hydrogen vehicles on the roads. By way of comparison, when EPACT was enacted in 1992 to promote alternative fuel use, there were already over 20,000 CNG vehicles and over 200,000 LPG vehicles operating in the United States (Energy Information Administration, 1996).

Furthermore, while most other alternative fuels were considered only for internal combustion engines, hydrogen's best use is seen in fuel cell vehicles. Although they offer great potential, fuel cell vehicles are the subject of intensive research and development. Thus, we conclude that even if hydrogen is more than just another alternative fuel because of its unique attributes, the same obstacles that faced alternative fuels, and more, may stand in the way of a successful transition to a hydrogen future.

Similarities Between the Hydrogen Transition and Other Alternative Fuel Programs

The hydrogen policy context may ultimately be different than it was for alternative fuels given the greater emphasis being placed on simultaneous reduction in transportation oil use and greenhouse gas emissions as the explicit raison d'être for the hydrogen initiative. Hydrogen fuel and fuel cell vehicles do offer the potential to take vehicles out of the energy and environmental equation. However, this has yet to play out in terms of sustained commitment and serious market forcing policies. That is, until the policy context changes, the hydrogen transition will be governed by the same market and policy conditions that governed the introduction of other alternative fuels.

Thus, while the policy context for hydrogen could be different in the future, it isn't yet. Many of the same approaches used to develop, promote, and deploy alternative fuel vehicles (AFVs) are now being used to develop, promote, and deploy hydrogen fuel and fuel cell vehicles. For example, the following were key ingredients in early and continuing efforts to expand the use of alternative fuels:

- Research and development
- Demonstration projects
- Fleet deployment
- Niche market development
- Public–private partnerships

Until the policy context changes, the best clues to what the hydrogen transition will look like may be in the not-so-distant past. So, let us look at what happened and what we can learn from the last 15 years of largely unsuccessful efforts to bring meaningful amounts of alternative transportation fuels into use.

Key Policy Initiatives in the Pursuit of Alternative Fuels

Figure 12-1 lists the critical federal laws enacted in the past 15 years to promote alternative fuel use to displace oil and reduce transportation air pollution in the United States. A bipartisan interest in alternative fuels in Washington, DC, led first to the Alternative Motor Fuels Act of 1988 (AMFA).

1988 - Alternative Motor Fuels Act
1990 - Clean Air Act Amendments
1991 - Executive Order 12759
1992 - Energy Policy Act
1993 - Executive Order 12844
1996 - Executive Order 13031
2000 - Executive Order 13149

FIGURE 12-1. Federal legislation promoting alternative transportation fuels.

It was comprehensive, including requirements for the DOE to conduct a series of analyses on alternative fuels, establish a research and development program, conduct studies, and create an interagency Commission on Alternative Motor Fuels. In addition, the Department of Transportation was directed to modify the corporate average fuel economy (CAFE) regulations to establish extra fuel economy credits for the manufacture of AFVs. CAFE credits became available for both dedicated and dual fuel AFVs, a controversial approach that would have a large impact on the future alternative fuel vehicle industry, but not the fuels industry. At that time, the primary alternative fuels under consideration were natural gas, methanol, and battery electric. AMFA was similar to a variety of state government programs, including the California program that aggressively pursued alternative fuels.

Soon after AMFA became law, a broad debate began in Washington, DC, to amend the 1970 Clean Air Act. Environmental advocates and alternative fuel proponents frequently endorsed different approaches, but even so a major alternative fuel initiative emerged. The Clean Air Act Amendments of 1990 established federal regulatory mandates for cleaner conventional fuels and vehicles, and created a new category of motor fuel, called reformulated gasoline, which was required to include oxygenated additives to reduce pollution. The potential environmental benefits of AFVs, including reduced emissions of ozone precursors, were used to "raise the bar" for conventional fuels and vehicles. Alternative fuel advocates began to learn that environmental benefits were a stronger motivator for introduction of alternative fuels than oil displacement.

During the days before the Iraq war of 1991, energy policy emerged again as a bipartisan priority. President George Bush issued Executive Order 12844, expanding on the federal fleet requirements of AMFA. Then, in the fall of 1992, the Energy Policy Act of 1992 became law, establishing broad goals and multiple new requirements for increased use of alternative fuels. The focus on fleets continued with new regulatory requirements for federal agencies, state governments, and fuel providers to purchase AFVs. Grants to state and local governments were authorized, educational and voluntary programs were established, and research and development programs were expanded.

During these same years, various tax credits were established to provide incentives for the use of battery electric vehicles and alternative fuels. Although many of the alternative fuels received some incentive, the biggest beneficiary by far was ethanol. During the 1990s, ethanol producers did not have to pay $0.54 per gallon of ethanol for road taxes when ethanol was used as a transportation fuel.

What Did the United States Get for This Effort?

The United States has conducted a broad and expanding alternative fuel program since 1988. What has been the result? Our review of the current transportation system in the United States suggests five direct and indirect outcomes:

- No significant change in alternative fuel use
- Cleaner conventional fuels and significantly lower vehicle emissions
- Expanded oxygenate fuel use
- Millions of alternative fuel compatible vehicles on the road, dominated by ethanol compatible vehicles
- Better understanding of alternative and conventional fuel markets, and consumer–producer behavior

The rapid growth of alternative fuel consumption promoted by stakeholders and endorsed by Congress simply did not materialize despite significant financial and policy investments. The Energy Policy Act of 1992 established a goal for alternative fuel use of 10 percent by the year 2000 and 30 percent by the year 2010, as a percentage of light duty motor fuel consumption. These aggressive goals would have required sales of AFVs to reach more than 30 percent of all motor vehicle sales by 2000, or about 5 million vehicles per year (Fig. 12-2). It also would have required investments of billions of dollars in refueling infrastructure.

Instead, the consumption of alternative fuels in AFVs increased only gradually, rising to less than 1 percent during the late 1990s. Even though AFVs have been increasing in number by about 7 percent per year, a much higher rate than overall vehicle population growth of about 1.5 percent, the starting point and growth rate are much too low to cause significant market

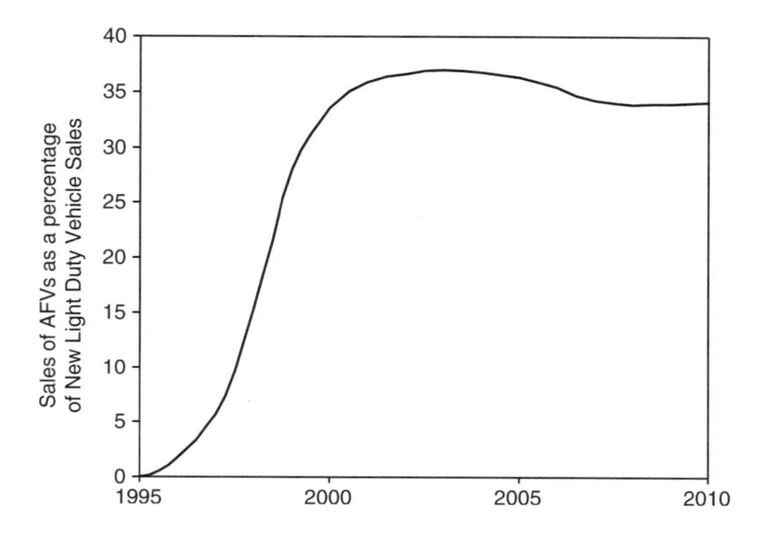

FIGURE 12-2. Percent sales of AFVs needed to meet EPACT goals.

penetration. Refueling stations expanded dramatically on a percentage basis but barely made a ripple compared to the number of gasoline and diesel stations. Most of this consumption was due to propane and natural gas vehicles that already had a head start. If displacing petroleum with alternative fuels was our only goal, these efforts were a failure.

The trends in lower vehicle emissions were more dramatic than fuel displacement and more sustained. Reformulated gasoline in California, and later at the federal level, provided significant reductions in ozone precursors when used in existing vehicles. Tougher tailpipe standards for new vehicles began providing emission reductions in new vehicles soon after the Clean Air Act Amendment of 1991, and continued through the decade. AFVs often led the way in emissions reductions, for example, when natural gas vehicles were the first to meet the federal inherently low emission vehicle standard and California's ultra low emission vehicle standard. Battery electric vehicles, of course, were the explicit goal for the zero emission vehicle (ZEV) standards in California.

The significant per vehicle emissions reduction of alternative fuels was soon eclipsed by the huge emission benefits coming from millions of gasoline vehicles that kept getting cleaner at the margin. At the end of the decade, emissions control advances had all but eliminated the emissions advantages of alternative fuels in light duty vehicles, though benefits in heavy duty AFVs still remained. There is no doubt that promotion of alternative fuel technologies made it possible to toughen regulations for conventional fuel vehicles, and if reducing emissions was the only goal, we were stunningly successful.

As a direct result of the Clean Air Act Amendments and the tax subsidy for ethanol, the use of nonpetroleum blending stocks in gasoline increased dramatically. Methyl tertiary butyl ether (MTBE) consumption increased from less than 1 percent to over 4 percent by 1995, primarily in reformulated gasoline. Ethanol use as a blending component in conventional gasoline increased from 0.5 percent to about 2 percent by the end of the decade. The multiple advantages of these blend stocks, including higher octane, low sulfur, and in most cases, decreased emissions, made them the logical picks for meeting the multiple regulatory requirements facing the fuels industry.

Indeed, the methanol industry all but abandoned support for the methanol fuel vehicle market it helped launch in 1988, as demand for MTBE consumed most of the world supply for methanol needed to produce it. At the end of the decade, the use of MTBE began to decline over concerns about water quality impacts, but ethanol use continues to grow at a steady pace. If making a market for agricultural products was a goal, we are increasingly successful.

The lack of alternative fuel refueling infrastructure created a "chicken and egg" problem for automotive manufacturers. Volume production of AFVs could not begin unless infrastructure was available, but the infrastructure investments would not be made unless numerous AFVs were being manufactured and sold. An innovative solution in the early years of AFVs was the dual-fuel vehicle. In the case of propane and natural gas vehicles, which were often aftermarket conversions, the vehicle would have two separate fuel systems—one for gasoline and one for the alternative fuel. The vehicle operator could switch to gasoline when the alternative fuel was not available.

Alcohol fuel vehicles were also capable of dual-fuel operation through the addition of an oxygen sensor that could adjust engine operations to any combination of ethanol or methanol with gasoline. Since alcohol dual-fuel vehicles had just one tank, they were dubbed flexible-fuel vehicles (FFVs). Theoretically, the large scale production of dual-fuel vehicles would allow motorists the security of gasoline fuel while creating a nascent demand for alternative fuel that would attract investment in refueling infrastructure. Encouragement of dual-fuel vehicles was considered important enough that legislators provided incentives under the CAFE requirements of the Energy Policy and Conservation Act of 1988. Some environmental stakeholders opposed credits for dual-fuel vehicles, saying the vehicles provided little benefit. As it turned out, they may have been right. During the 1990s, the sale of dual-fuel vehicles expanded as anticipated. Several manufacturers offered dual-fuel natural gas vehicles that were among the bigger sellers to fleets. Methanol and ethanol FFVs were also offered, and soon became the lowest cost option to obtain an AFV.

During the mid-1990s, production of methanol fuel vehicles was phased out, and ethanol FFVs began to increase. Designed to run on E85, an 85 percent blend of ethanol and 15 percent gasoline, almost all these vehicles were almost always fueled with gasoline because E85 stations were few and far between and ethanol fuel cost considerably more than gasoline, even with subsidies. Production of FFVs soon far outstripped the demand for AFVs as automotive manufacturers took advantage of the CAFE credits to help meet car and light truck CAFE standards. By the end of the decade, well over three million ethanol FFVs had been manufactured. In the end, it appears the dual-fuel experiment did not meet the expectations of the promoters. Despite the existence of millions of FFVs, which do provide a large nascent demand for ethanol in the form of E85, the growth of E85 stations has not kept pace despite efforts by producers, primarily in the Midwest.

If creating a latent market for neat ethanol (E85) use was a goal, the United States has succeeded at least in part. If another oil crisis occurred, the E85 vehicles could theoretically avoid gas lines by stopping at the local E85 pump. If there were enough FFVs and enough E85 pumps, and a way to rapidly expand ethanol production in a crisis, this would enhance energy security. But the marginal benefits are unclear in that ethanol can be readily used as a gasoline blend stock at the 10 percent level in conventional vehicles, thus providing a ready avenue for distribution of all available ethanol to the largest number of vehicle users in a crisis.

If improving the analytical basis for developing and implementing future policies was a goal, the United States has been quite successful. Beginning with AMFA, a variety of studies, reports, analyses, and models were developed to increase understanding of alternative fuels and vehicles. Many of these were funded by the federal DOE and performed at its national laboratories, but state governments and academe were also actively involved. Lifecycle analysis, called well-to-wheel analysis when applied to transportation systems, became widely used and is now considered fundamental to understanding the impacts of the transportation sector on energy and the environment. Analyses and debate over alternative fuels became a regular feature at meetings of the Transportation Research Board, part of the National Academy of Sciences. The result has been a tremendous increase in the understanding of how transportation fuel markets operate and how consumers and producers of fuels and vehicles behave in those markets. Most of what has been learned is directly transferable to understanding the issues for a transition to hydrogen fueled vehicles. This increased understanding was not limited just to alternative fuels. Analytical work covered areas such as technical information on fuel costs, environmental impacts, vehicle technology, market economics, world resource bases, oil markets, and more. If advancing the technical knowledge base was a goal, the United States succeeded.

Lessons Learned

One of the key lessons learned from U.S. transportation energy programs of the last 15 years is that entrenched conventional automotive technology is hard to displace. Alternative fuel advocates made a convincing case in the late 1980s and early 1990s that vehicles operating on alternative fuels were technically superior to those using petroleum products and could help reduce emissions and reliance on foreign sources of energy. With several legislative initiatives providing policy drivers, and state and federal governments promoting alternative fuel use through regulations, many assumed that momentum for alternative fuels would develop and lead to significant displacement of petroleum before the year 2000.

What happened instead is a lesson in competitive response. Industries with significant sunk investments in conventional transportation fuels and conventional vehicle technologies were able to identify and implement significant technical improvements that began to weaken the policy argument for alternative fuels. Reformulated fuels were able to deliver significant emissions reductions, not only in new vehicles but in the existing vehicle stock, that dwarfed the potential that alternative fuels could provide in the short term. Advances in conventional and advanced vehicle technologies delivered both cleaner and more efficient vehicles. For example, by the time the Environmental Protection Agency (EPA) had promulgated the rule for the Clean Fuel Fleet program in 1993 that was originally thought of as a virtual mandate for AFVs, the automobile manufacturers had voluntarily agreed to the national low emission vehicle program that delivered cleaner vehicles to more people. The lesson learned here is that new technologies have to be better, in a meaningful way, not just different, and must keep ahead of conventional technology improvements that will inevitably occur. Time and again, alternative fuel advocates were able to claim credit for "forcing" conventional technology and fuel providers to improve, but this was little comfort to the companies that made and in most cases lost their financial investments in alternative fuels.

Another lesson learned is that niche markets don't necessarily grow into mainstream markets for light duty vehicles, even with significant government support. The authors of AMFA and EPACT hoped that federal demonstrations and fleet requirements, coupled with education, outreach, and some small incentives, would lead to widespread use of alternative fuels in the light duty fleet. Early success in small or niche markets, it was theorized, would catalyze additional demand for AFVs, giving manufacturers confidence to expand offerings and increase supply. Fueling infrastructure built to serve the niche market "early adopters" would be expanded to serve a growing consumer demand, or so it was thought.

In the early 1990s, government and industry partners worked to identify the types of vehicles most likely to generate early demand in these niche markets and tried to consolidate demand for available vehicles to encourage economies of scale. For a variety of reasons, this commonly endorsed approach had little success. Vehicle models used by fleets did not match the needs of general consumers. Trade-offs acceptable to niche markets, such as limited range, were not acceptable to general consumers. Vehicle design changes to accommodate gaseous refueling systems or electric drive systems were too expensive for manufacturers to extend to additional models. Limited budgets and engineering staff at manufacturers and government agencies meant that a small subset of all alternative fuel combinations even reached the stage for emissions testing and certification. Centralized refueling infrastructure to support early adopters, such as transit agencies, was often not available to general consumers.

Thus, there is no evidence that alternative fuel use in niche markets leads to expanded mainstream markets in light duty vehicles. Somewhat greater success with alternative fuels in heavy duty vehicle niche markets, such as garbage trucks, school buses, and utility trucks, has convinced some alternative fuel proponents that niche market development can work in certain situations.

As a corollary to the observations about niche markets, it also appears that vehicle fleets are a bad place to try uneconomical fuels. Fleets were originally considered one of the best niche markets to become "early adopters" of alternative fuels. Fleets, it was assumed, had centralized refueling, centralized purchasing, and cost structures that would allow them to absorb the incremental cost of adopting alternative fuels and vehicles.

Yet virtually none of the assumptions about fleets turned out to be correct. Most light duty fleets are no longer centrally refueled but rely on conventional retailers. Most fleets have a primary goal of achieving low operating costs, unlike general consumers, and were very resistant to alternative fuels and vehicles that carried high incremental cost. Even fuels such as natural gas and propane that could offer lower lifecycle costs had difficulty selling to fleets that placed a premium on first costs. Fleets that are centrally fueled, such as transit districts, often buy fuel in bulk at attractive prices, making alternative fuels even less attractive. Most fleets have longstanding relationships with vehicle and engine suppliers that are not easily dissolved. Many fleets turn over vehicles quickly, relying in part on high resale value to help lower lifecycle costs. The lack of a viable resale market for AFVs made fleets resistant to their adoption. For these and other reasons, fleets turned out to be poor early adopters (Nesbitt and Sperling, 1998).

Turning to fuel quality issues, the experience of the last 15 years suggests that low energy density fuels have high consumer refueling costs that affect consumer choice. Authors of AMFA, EPACT, and various state government requirements assumed that alternative fuels could be delivered to a vehicle in pretty much the same way as conventional fuels, and that any concerns, such as refueling or limited range, could be overcome with some training and education. Indeed, where alternative fuels have been successful, such as transit buses, education and training have been critical to overcome barriers to introduction.

Yet, no amount of education seems to be able to make up for the fact that low density fuels that require more frequent refueling impose real costs on fleets and consumers. For example, a fuel like CNG or LPG can be delivered to the vehicle for less than the price of gasoline in most of the country, but for many fleets, labor costs for extra driving time to find an alternative fuel refueling station, and the extra time for refueling, can swamp all other potential cost savings, especially in vehicles with low annual mileage. For consumers, more frequent refueling coupled with extended search times are estimated to contribute as much as $1.00 per gallon to the effective consumer costs for alternative fuels in light duty vehicles (see Chapter 14 of this volume).

Another lesson stems from the big differences between advocates or special interest groups and the response of the consumer markets. Alternative fuel advocates tend to emphasize the positive attributes of their fuel, while minimizing the negative attributes. In the policy arena, the positive attributes helped sway decision makers to adopt programs to promote alternative fuels. In the marketplace, the negative attributes have, so far, limited alternative fuel use to a small number of applications. Any future alternative fuel program must try to better understand what negative attributes of the new fuel and vehicle combination need to be addressed proactively before deployment begins. Persuasive advocates are needed to help build effective public policy and launch alternative fuel programs, but functioning markets and satisfied consumers are required for any successful alternative fuel effort.

Given consumer reticence, the political system has not yet shown a willingness to impose significant visible costs on private players in the name of the fuel diversity promised by alternative fuels. A review of worldwide alternative fuel policies has shown that public policies can be effective in expanding and maintaining the use of alternative fuels, such as in Brazil and Denmark (U.S. Department of Energy, 2000). In the United States, however, the policies on the books are not sufficient to provide disincentives to the use of conventional fuels or incentives to the use of alternative fuels for a significant number of vehicle users. This is despite the momentum developed with the passage of federal legislation over a 4-year period and the ensuing 15 years of effort at the federal and state level to use these legislative drivers to promote alternative fuels. The fuel and vehicle incentives on the books are not adequate to defray the higher incremental cost of existing alternative fuels and vehicles. The regulatory support for alternative fuels, limited in the first place, is further weakened by changes in technology and the marketplace, and is not adequate to induce significant numbers of vehicle purchases.

Unregulated and unsubsidized private sector investment in refueling infrastructure has proven to be very limited. The growth in alternative fuel refueling infrastructure has been simultaneously dramatic and disappointing. Refueling infrastructure has grown for each of the high visibility alternative fuels except methanol, yet most of these stations were established in whole or in part with state and federal government funding. In particular, many natural gas refueling and electric vehicle charging stations were built between 1990 and 1995 by regulated utilities with the permission of public utility commissions (PUCs). As these industries entered a period of deregulation, the PUCs reduced and then eliminated these resources, and station development slowed significantly. The most successful and sustainable infrastructure development has been centered on so-called "anchor fleets" that can provide fuel demand sufficient to provide a return on investment. These anchor fleets tend to be heavy duty, high fuel consumption fleets. One of the more notable successes, the Interstate Clean Transportation Corridor, was organized in California to support delivery trucks (U.S. Department of Energy, 2003).

Building infrastructure in anticipation of market development has rarely happened, and when it has, the investors have usually been disappointed, especially with high cost refueling stations, such as for natural gas or electricity. Fuels with lower infrastructure investment costs, such as ethanol, would appear to have lower barriers to introduction. However, the target markets for these fuels are general consumer vehicles which refuel at branded and independent stations. Branded stations have been reluctant to offer fuels not supported by the parent companies; both types of stations have thin profit margins that increase the opportunity costs of replacing a conventional pump with an alternative fuel.

Unfortunately, consumers are reluctant to accept obvious trade-offs even when the private system can justify investments in AFVs and alternative fuel refueling infrastructure. Most vehicle consumers want it all. They want a low emissions vehicle, but insist that it be powerful as well. They want their large, heavy vehicle to be fuel efficient. They won't give up trunk space for a large gaseous fuel tank. They don't want to refuel on the opposite side of town, or even the opposite side of the street, even if that's where the alternative fuel is located. They want the reduced emissions and low noise of alternative fuel transit buses, but they also want low fares.

Surveys indicate that consumers favor government mandates or taxes to reduce oil dependency even while individual consumer choices result in the opposite (U.S. Department of Energy, 2001). The political system has reflected this by limiting regulatory requirements on the automobile and fuel industries to emission reductions and safety improvements that are largely transparent to the end consumer. California's controversial zero emission vehicle program was criticized, in part, because it required manufacturers to sell ZEVs but did not have a parallel requirement for consumers to buy them.

Social attributes of the new alternative fuels are not valued by mainstream consumers. Consumers have demonstrated the capability of being supportive of energy and environmental solutions while simultaneously purchasing vehicles that move in the opposite direction (e.g., toward higher horsepower). Fleets, with their emphasis on reduced operating costs, are often resistant to paying extra for societal benefits. Even when some consumers are willing to pay extra to be environmental leaders or early adopters, the programs designed to motivate and encourage these consumers don't necessarily attract the general consumer. Too few consumers, for example, demonstrated a willingness to accept a vehicle with very low range, even if it had zero emissions and used no oil. If the new fuel or vehicle technology is not better for some recognized and valued performance criterion, they will not buy it.

Another lesson learned at the DOE over the past 15 years is that coordination between auto and oil industries is vital. This should not be surprising, but it can be a significant limiting factor in developing successful joint efforts. Experience with conventional and alternative fuels indicates that coordination can sometimes be hard and is often driven by self-serving views and needs. For example, the composition and distribution of federal reformulated gasoline, required by the Clean Air Act Amendments of 1990, had a significant impact on the ability of automakers to implement emissions reduction technologies. At the end of the 1990s, debates over low-sulfur diesel fuel illustrated the competing needs of auto and oil industries. In the area of alternative fuels, vehicle manufacturers had to establish partnerships with nontraditional energy organizations, such as the Gas Research Institute, recently renamed the Gas Technology Institute, to accelerate the development of technology, codes, and standards.

Government organizations often played a vital role in facilitating these new partnerships.

It seems, based on experience to date, that transition technologies often become so good as to overwhelm the end point technology. This could be an especially critical factor influencing the pace and direction of the hydrogen transition that lies ahead. As an example from the past, the low cost and low technology hurdle of FFVs, coupled with policy incentives for manufacturers, led to the dominance of FFVs over all other types of AFVs in the light duty market. Similarly, reformulated gasoline was first conceptualized as a way to fight back against the emissions performance of methanol vehicles, but its widespread use totally eclipsed the potential penetration of alternative fuels, including methanol, in just a few years. Finally, hybrid electric vehicle (HEV) technologies provided many of the benefits of battery electric vehicles, but without the high cost and range limitation. This helped convince many that ZEVs were no longer needed. Marketing campaigns for HEVs even played on consumer concerns about battery electric vehicles. Consumers were promised that, as an HEV owner, "You don't have to plug it in."

A final lesson from the past 15 years is that size and scale matter in a significant way in vehicle and fuel markets. Policymakers originally believed that use of fleets and other niche markets could catalyze the widespread adoption of alternative fuels and vehicles in the consumer market, but this has not proven to be true.

The very size and scale of the light duty market makes it difficult for alternative fuels and vehicles to break in. Significant penetration in the light duty consumer market is required for significant oil displacement because the light duty market accounts for 75 to 80 percent of U.S. transportation oil consumption. In the United States, 15 to 18 million light duty vehicles are purchased each year; light duty fuel consumption is well over 100 billion gallons per year. Alternative fuels must compete on price with conventional fuels that have achieved huge volumes of production with access to an established low-cost distribution infrastructure and a ubiquitous delivery system of refueling stations.

Attracting investors willing to challenge the established transportation fuels market is difficult. Vehicle manufacturers have little incentive to launch expensive engineering efforts for AFVs that will reach sales of less than 1000 or 10,000 per year. Most manufacturers cannot make a profit on demand of less than 50,000 per year per model. Policymakers and regulators eyed the federal government fleet of 450,000 vehicles as a prime candidate to help launch AFVs. Yet many of these vehicles are in the Defense Department, many with tactical requirements, and many more are in the U.S. Postal Service, which aims to replace vehicles only every 25 years. Federal agencies do purchase approximately 45,000 new light duty vehicles every year, which looks at first blush to be a sizeable market. But this total is divided among 10 to 15 different models and size classes. Unlike other new technologies, such as efficient lighting, where federal government demand can exert a substantial influence on total U.S. demand for a product, federal fleet purchases of AFVs have proven to have almost no influence on manufacturer decisions.

AFV penetration has been most successful in markets that are small, well-defined, and with limited competition, such as in the transit bus, garbage truck, school bus, and taxi markets. Yet these are the same markets least likely to catalyze broader market penetration. Some have suggested alternative fuels, especially battery electric vehicles, may be able to carve out new market niches as "disruptive technologies" that serve unmet consumer needs (Christensen, 1997). Yet the size, scale, and diversity of the existing transportation market works against this by providing a huge portfolio of options for almost every consumer transportation need.

Conclusion

What, if anything, does the 15-year experience with alternative fuels, and the lessons learned, tell us about the upcoming hydrogen transition? We think the lessons learned should be studied in detail by those who have the most to gain from a successful introduction of hydrogen as a transportation fuel, and also by those who have the most to lose from less than full success. We believe the numerous analyses and models developed to assess alternative fuels should now be harnessed to help chart the most effective course toward a hydrogen future. In particular, we have made five overarching observations drawn from the experience with alternative fuels in the United States to date.

First, lower energy density fuels significantly raise the cost of infrastructure and impose ongoing "time" costs on consumers. Competitive range, and speedy and simple refueling may be the most critical technical barrier to broad consumer acceptance.

Second, there is no evidence that niche market activity will lead to expansion into mainstream consumer markets. Hydrogen and fuel cell vehicles will have to have an early and significant presence in mainstream consumer markets to serve as a basis for a successful transition.

Third, the incremental benefits of hydrogen and fuel cell vehicles to consumers are likely to be very small relative to conventional technology vehicles and fuels. Policies to value the social benefits of hydrogen and fuel cell vehicles will be needed, or at least considered, to stimulate consumer demand enough to achieve the timetable and scale of transition envisioned.

Fourth, infrastructure development may be the limiting factor. It is hard, and perhaps impossible, to see how private sector infrastructure investment will be made in the timeframe and scale needed to achieve success without clear stimulus from government. This is true even assuming hydrogen technology performance and cost goals are achieved. High risk and costly investments are needed to overcome the chicken and egg issue.

A national policy framework to bring about fuel infrastructure development will almost certainly have to be part of the longer term plan.

Finally, the broader policy context, which values the attributes of alternative fuels, including hydrogen, is still missing in the U.S. energy debate. Instead, the debate focuses on pushing technology development and deployment for fuels and vehicles using indirect tools. This has not been successful over the past 15 years of alternative fuel efforts.

Given these five critical observations, the emphasis in the U.S. hydrogen initiative on successfully addressing the technology barriers to hydrogen and fuel cell vehicles through research and development seems right on the mark. The role of research and development as an element of programs to promote alternative fuels has varied widely. While work during the past 15 years has advanced the state of the art for alternative fuels and AFVs, R&D did not and was not generally expected to fundamentally change the cost, performance, or market acceptance of these fuels and vehicles, except perhaps for battery electric vehicles. Clearly the situation with hydrogen and fuel cell vehicles is different in that a significant amount of government-led long term and high risk research is needed to reduce the cost and improve competitiveness of hydrogen and as such is an appropriate policy response at this time.

Acknowledgments

Thanks to Tom White for his help in preparing the Asilomar presentation and to Richard Bechtold and Melissa Lot for their review of and suggestions for the paper.

References

Cannon, James. 1995. *Harnessing Hydrogen: The Key to Sustainable Transportation.* New York: Inform, Inc.

Christensen, Clayton. 1997. *The Innovator's Dilemma: When New Technologies Cause Great Firms to Fail.* Cambridge, MA: Harvard Business School Press.

Energy Information Administration. 1996. "Alternatives to Traditional Transportation Fuels 1994." Vol. 1. Washington, DC: Energy Information Administration, February 1996.

Nesbitt, K., and D. Sperling, 1998. Myths Regarding Alternative Fuel Vehicle Demand by Light-Duty Vehicle Fleets. *Transportation Research* D. 3(4): 259–269.

U.S. Department of Energy. 2000. "Replacement Fuel and Alternative Fuel Vehicle Technical and Policy Analysis." Washington, DC: U.S. Department of Energy, September 2000, p. 49.

U.S. Department of Energy. 2001. "Fact of the Week #162." http://www.ott.doe.gov/facts/archives/fotw162.shtml. Golden, CO: U.S. Department of Energy.

U.S. Department of Energy. 2003. http://www.ccities.doe.gov/corridor/interstate_corridor.shtml.

CHAPTER 13

Lessons Learned in the Deployment of Alternative Fueled Vehicles

Bernard I. Robertson and Loren K. Beard

The United States (U.S.) is in the process of embracing the hydrogen powered fuel cell vehicle as the answer to the nation's transportation needs, at least for the second half of the twenty-first century, and perhaps before then as well. The scientific, engineering, social, consumer acceptance, economic, and policy hurdles that must be overcome to enable this technology to succeed are unprecedented. This chapter analyzes past deployments of alternative fuel vehicles (AFVs) by DaimlerChrysler Corporation to reveal the lessons learned so that potential stumbling blocks or inconsistent policies can be avoided in the deployment of hydrogen powered cars and light duty trucks.

A "Holistic" Approach to a National Energy Policy

A primary consideration in the decision to deploy AFVs is that the justification and plan for the deployment of these vehicles must be part of an integrated national energy policy, which includes not only stationary sources, but also the entire range of energy use in the United States.

The first element of a national energy policy should be to identify an overall objective. Is the national objective to limit emissions of greenhouse gases? Is it to reduce dependence on petroleum and other fossil fuels because the resources are finite and future supplies are questionable? Is the national objective to reduce the use of imported petroleum, especially that petroleum which is imported from politically unstable or even hostile regions? A clear, precise statement of this energy goal is necessary so that proposed programs can be evaluated as to their contribution to that objective. Absent this step, multiple solution sets must be pursued, at great cost, with the inevitability that much of this effort will be wasted.

Once energy policy objectives are clearly articulated, appropriate achievable and quantifiable goals and targets must be set. Is our goal to stabilize atmospheric carbon dioxide (CO_2) levels at 500 parts per million (ppm), or is the more ambitious goal of 300 ppm or preindustrial levels necessary? What level of imported raw energy resource imports is acceptable to provide for national security? Is it 50, 30, or 0 percent? What segment of the economy offers the most cost effective opportunities for reductions in energy use? Could it be the industrial stationary sector, or is the residential or transportation sector more appropriate? Should the effort focus on industrial operations or turn to consumers for leadership? How should all these parameters be combined to form clear targets for energy policies?

An energy policy must focus on the total lifecycle energy use or greenhouse gas emissions of a system. Such lifecycle analyses must assess the raw energy resource utilized, whether in the form of petroleum, natural gas, coal, nuclear, solar, hydro, or wind. The energy expended in processing the raw energy resource into a usable fuel and providing that fuel to consumers in a usable form must then be calculated. For transportation analyses, the energy used in the manufacture, transport, and eventual disposal of all components of the vehicle or other system must be known, in addition to the energy utilized by the vehicle or system during its functional life. Finally, externalities, such as positive or negative impacts on local environments, societal impacts, and resources needed to secure the availability of the raw energy resource, should be considered in any lifecycle analysis.

An energy policy must contemplate and accommodate market realities. An AFV must provide all the customer-desired attributes of conventionally fueled vehicles at a competitive cost, and with minimal inconveniences, such as prolonged refueling time, widespread availability of fuel, and other factors. Financial incentives from governments can be helpful in the introductory phases of an alternative fuel program, but they must have a finite lifetime, and the technology must compete on an even footing in the marketplace if it is to achieve widespread market penetration.

Chrysler's Experience in the Deployment of AFVs

Since the late 1980s and long before its merger with Daimler Benz to form DaimlerChrysler AG, Chrysler has been developing a wide range of AFVs for sale in the U.S. market. Light duty vehicles fueled primarily by natural gas, methanol, and ethanol have all been offered for sale by Chrysler. When electric vehicle technology emerged, Chrysler also offered a battery powered minivan, and, more recently, it has built a number of prototype fuel cell electric vehicles. The history of these vehicle introductions is summarized below. The products discussed here were developed and deployed by Chrysler Corporation or the Chrysler Group of DaimlerChrysler. In either case, they will be referred to as Chrysler products.

Compressed Natural Gas (CNG)

Between 1992 and 1996, and again between 1998 and 2001, Chrysler produced over 6000 pickup trucks, minivans, and full-sized vans capable of burning CNG as their sole fuel. The trucks were equipped with a natural gas engine derived from a 5.2-liter, V-8 gasoline engine, and minivans with a natural gas engine derived from a 3.3-liter, V-6 gasoline engine. All were dedicated CNG vehicles, no longer capable of burning gasoline. Chrysler chose to produce dedicated CNG vehicles, rather than bifuel vehicles capable of burning either natural gas or gasoline, to optimize the environmental performance and avoid some of the compromises in packaging, performance, cost, weight, and emissions that are difficult or impossible to avoid in a bifueled vehicle.

The CNG vehicles were a technical success and included the first vehicles certified to the strict California low emissions vehicle (LEV) and ultra low emissions vehicle (ULEV) standards. They were also certified to the inherently low emissions vehicle (ILEV) category, and the full-sized van was certified to the California super ultra low emissions vehicle (SULEV) standard.

The objectives of the CNG program were to produce a generation of cleaner burning vehicles, particularly for applications in congested areas such as mass transit and airport shuttle activity, and to displace petroleum usage by natural gas, which was more readily available from domestic sources. Incentives and mandates were provided to help bring these vehicles to market. While technically successful, the vehicles didn't enjoy widespread success. The CNG program never advanced past the dedicated fleet niche, in part because the fleet mandates enacted in local and state air pollution programs or contained in the federal Energy Policy Act (EPACT) of 1992 were never fully enforced.

Furthermore, retail incentives were minimal. Consequently, refueling sites for natural gas vehicles were largely limited to private fleet sites, and a widespread public refueling infrastructure has not developed. The absence of such a public refueling infrastructure has limited the secondary market for the vehicles and has consigned them to a niche market where they continue to provide a very clean alternative for shuttle, package delivery, and other fleets. Since fleet operators rely heavily on lifecycle analysis, the absence of a secondary market has been a major obstacle. Finally, a pall was cast over the natural gas vehicle business after two high pressure tank failures at 3000 pounds per square inch (psi, or 210 bar) on a competitor's vehicles led to that company repurchasing the vehicles and exiting the market.

M85

M85, a solution of 85 percent methanol with 15 percent gasoline, was a transportation fuel that attracted great interest in the early 1990s.

Experiments had shown that internal combustion engines (ICE) running on methanol could achieve much lower emissions levels than contemporary gasoline engines. Thus, the reduction of local emissions was a key driving force. There was also an opportunity to displace petroleum usage with methanol, which was largely derived from natural gas, a plentiful and domestically available fuel.

Methanol faced two major problems: Pure or "neat" methanol had physical characteristics that make its use in motor vehicles problematic. Its very low volatility made it difficult to start the vehicle at low temperatures, and its nearly invisible flame raised safety concerns in the event of a collision or other spillage. These problems were solved by blending 15 percent gasoline into neat methanol to produce M85, which had sufficient volatility for cold starts and burned with a pronounced gasoline-like orange flame.

Another problem came to be known as the "chicken and egg" quandary. Automakers were reluctant to build vehicles for which few if any refueling stations were available, and the fuel industry was reluctant to build refueling stations for vehicles that would initially be built in the thousands, not in the millions. A very elegant solution to this problem was developed for methanol AFVs, and known as the flexible-fueled vehicle or FFV. These vehicles were designed to run on M85, gasoline, or any mixture of the two. Thus, automakers could introduce the vehicles knowing that if M85 was not conveniently available, the customer could run on conventional gasoline until another source of M85 could be obtained. In principle, once sufficient numbers of FFVs were on the road, incentives and subsidies, coupled to government fleet mandates, would lead to the development of a broad M85 refueling infrastructure.

The vehicles were a technical success, and thousands were built and sold in the early to mid-1990s. Chrysler, for example, sold 11,911 FFV passenger cars, mostly in California. However, despite subsidies, a general refueling infrastructure for M85 never materialized, and Chrysler ceased marketing the vehicles in 1995. Most of our M85 vehicles are still on the road today, most of them burning gasoline exclusively.

Another issue with the M85 FFV was that while its principal focus was on the reduction of local pollutants, continuing improvement in conventional gasoline ICEs, coupled with the introduction of California Phase II reformulated gasoline in 1996, demonstrated that similar or stricter emissions performance could be achieved by gasoline powered vehicles.

E85

Building on lessons learned in the M85 program, in 1998 Chrysler launched the E85 minivan FFV capable of burning gasoline, a blend of 85 percent ethanol and 15 percent gasoline, or any combination of the two. The emphasis in this program was not reducing ozone forming local pollutants, as was the case for M85 vehicles, but the displacement of petroleum with ethanol, a renewable fuel. As an incentive to produce these vehicles, auto manufacturers were granted corporate average fuel economy (CAFE) credits under federal law.

Chrysler provided E85 FFV capability, at no cost to the customer, in all of its 3.3 liter, V-6 minivans between 1998 and 2002. More than a million of these vehicles were sold. Our competitors adopted similar strategies, and today there are nearly four million FFVs in service in the United States, all capable of running on E85. If a fuel infrastructure and appropriate incentives were in place to encourage these vehicles to run on E85, the United States could reduce its petroleum usage by over 1.5 billion gallons annually (or 100,000 barrels of oil per day) and avoid 7.5 million tons of CO_2 emissions per year. As more efficient processes for producing ethanol from other biomass resources are perfected, these benefits would increase significantly.

This program is ongoing, and new FFV offerings will be available in the future. The program had originally anticipated that E85 refueling stations would keep pace with the introduction of E85 FFVs. To date, however, only about 165 public refueling stations exist nationwide, so it is clear that more emphasis needs to be placed on the fuel infrastructure issues. In the meantime, incentives for the introduction of E85 FFVs should continue to help reach the critical mass of vehicles on the road, which will help spur fuel infrastructure development.

E85 FFV vehicles are transparent to the driver and offer significant environmental benefits. The key obstacle has been economics. The experience at Chrysler points up once again the difficulty of competing with cheap Middle East oil, regardless of the other attributes of the available AFVs.

Battery Powered Electric Vehicles (BEVs)

In the 1990s, in response to the California zero emission vehicle (ZEV) mandate, Chrysler invested very heavily in the development of a full function electric minivan, the EPIC (see Fig. 13-1). Given the limitations of battery technology, the vehicle was a technical success but had several problems associated with today's batteries. The BEV, for example, exhibited a very low range of under 100 miles, compared to more than 300 miles common in equivalent gasoline powered vehicles. The time for recharging was, and remains, a serious drawback. The lifetime of the battery package, especially given its cost, was also a major hurdle.

Because of the $120,000 cost of the vehicle, driven largely by battery cost, large subsidies were needed to lease the EPIC at rates comparable to its gasoline powered equivalents. Even when these heavy subsidies were offered, total sales over the life of the program reached only 281 BEVs.

While research and development in the battery electric arena was worthwhile, the leap to the California ZEV mandate was a classic case of a manufacturer sales mandate that did not adequately consider the state of technology, cost, customer demands, or refueling infrastructure inherent in

FIGURE 13-1. The EPIC battery electric minivan.

the vehicle technology. In terms of development of BEVs, the ZEV mandate was clearly a failure. The only bright spot was the growth in neighborhood electric vehicles (NEVs), which offer an environmentally attractive alternative to small ICE vehicles used for short trips. DaimlerChrysler has developed a successful business with the introduction of the GEM NEV (see Fig. 13-2). Those who claim that the ZEV mandate was successful point to the development of partial zero emission vehicles, or PZEVs, which represent the state of the art in gasoline ICE refinement. Though important, these are hardly a substitute for economically viable, full function BEVs, and in any event would probably be the next logical step in emissions control with or without the ZEV mandate.

Light Duty Diesel Vehicles

While most would not consider petroleum derived diesel fuel to be an alternative fuel, given that the U.S. light duty fleet consists of less than 1 percent diesels, diesel in the light duty sector is definitely an alternative to gasoline.

In the United States, the diesel engine has a reputation as a noisy, smelly, smoky, and sluggish power plant. These views are largely supported by passenger car diesels that were brought to market in the late 1970s and early 1980s in response to the beginning of the CAFE program, and to poorly maintained, old technology transit buses and freight haulers still operating in the United States today.

In reality, diesel engine technology has undergone a kind of revolution, not unlike that which took place in gasoline powered ICEs in the mid-1980s to mid-1990s, when open-loop carbureted systems were replaced by

FIGURE 13-2. The GEM neighborhood electric vehicle.

closed-loop electronically controlled, fuel injection systems. The diesels of the seventies and eighties relied on low pressure, mechanically controlled fuel injection, which gave the engine developer little control over the timing, duration, and droplet size of the fuel injection process. The advanced diesels of today utilize high pressure, electronically controlled, multiple direct injection, in combination with very precisely machined, multiple hole injectors, which allow for cycle to cycle control of the fuel injection event, and a much finer spray droplet size distribution. These qualities lead to dramatically reduced noise, vibration, and particulate matter emissions.

In fact in Europe, where customers place a higher value on fuel economy, diesels now account for about 40 percent of new vehicle sales. For the Chrysler Group of DaimlerChrysler, the number is closer to 66 percent. If Chrysler had the same diesel penetration in its U.S. fleet as it has in Europe, its CAFE performance would improve by about four miles per gallon. If the entire U.S. light duty fleet had the same diesel penetration as Europe, total U.S. fuel consumption could be reduced by about one million barrels per day, and CO_2 emissions could be reduced by almost 200 million tons per year.

In addition, slight material changes to diesel fuel systems, and a strict quality standard for biodiesel in the United States, could lead to substantially greater savings in petroleum usage and reductions in CO_2 emissions.

The U.S. Department of Energy, through the EPACT program, has designated B-20, a 20 percent mixture of biodiesel in conventional diesel fuel, as an alternative fuel, and many fleets are successfully operating diesel powered Dodge Ram diesel trucks on such fuels.

Another development in the diesel world is the growing interest in gas-to-liquid, or GTL, processes. Chrysler has worked with Syntroleum, a technology company that has developed a process to convert remote natural gas reserves now stranded from access to energy markets into sulfur free, aromatics free, high cetane diesel fuel that can be transported easily by tanker or pipeline to markets. This fuel has been shown to offer significant emissions decreases and customer satisfaction benefits. However, with crude oil hovering around the $26 to $30 per barrel level, the business case for the use of GTL fuels, like other alternative fuels, is difficult.

In Germany, the Mercedes Car Group has been working with Choren Energy on the development of a biomass-to-liquids (BTL) process, which produces high quality Fischer-Tropsch diesel fuel from wood waste. Initial results are promising, but again, absent initial government incentives, costs will be an issue.

A major challenge even for advanced diesel engines is the array of impending U.S. federal Tier 2 and California LEV II emissions standards. Work is ongoing to achieve these standards and to ensure that the exhaust after-treatment devices required will meet the necessary durability standards. Two things are certain: Diesel fuel quality must improve dramatically, at least to the 2006 ultra low sulfur diesel requirements and preferably well beyond, and the cost of a Tier 2 Bin 5 or LEV II compliant diesel system will be substantial, rivaling the cost of hybrid electric gasoline fueled vehicle systems.

Hydrogen Powered Fuel Cell Vehicles

The hydrogen powered fuel cell vehicle is seen by many as the power plant for future light and heavy-duty transportation, and even stationary sources. The fuel cell is often seen as a panacea because of its virtually zero criteria emissions, little or perhaps no CO_2 emissions, and complete independence from imported petroleum. However, in the light duty vehicle sector, the hydrogen powered fuel cell faces an array of enormous challenges.

Fuel cell technology is expensive. Current fuel cell system technology cost must be reduced by two orders of magnitude to approach the cost of today's ultraclean conventional ICE systems. In addition to cost, formidable challenges remain in cold weather start up and operation, cooling, durability, and onboard hydrogen storage.

Beyond the challenges presented by the development of a reliable, affordable fuel cell, the challenges associated with a refueling infrastructure are equally daunting. While there is a consensus that early fuel cell vehicle deployment will rely on onboard storage of compressed hydrogen, stored at pressures of 3000 psi and higher, DaimlerChrysler has also demonstrated direct methanol fuel cells, which derive their hydrogen from onboard methanol reformers. It is also studying a unique approach to storing hydrogen in sodium borohydride liquid, which can be hydrolyzed onboard to produce hydrogen gas and sodium borate. The sodium borate can be removed from the vehicle and recycled back into sodium borohydride.

Much work remains to be done to determine the most overall cost effective solution and establish the business case for deployment of a hydrogen infrastructure. At least in the early stages, the most likely candidate for the production of hydrogen would be from the steam reforming of methane at stationary sites off the fuel cell vehicle. Unfortunately, many of the potential CO_2 reduction benefits from hydrogen fuel cell vehicles would not be fully realized until renewable resources are developed.

DaimlerChrysler has begun the deployment of 30 compressed hydrogen fuel cell powered city buses and 60 compressed hydrogen F-cell vehicles derived from the Mercedes A-class. These vehicles are one step past the laboratory stage and are being used to evaluate their fitness for daily use. The field tests of these vehicles will assess their reliability and determine if they can be operated routinely in commercial settings without teams of engineers to maintain them.

While fuel cell vehicle market entry was accomplished in 2003, volume production is probably 10 or more years away, because of the challenges faced by the technology. Nevertheless, DaimlerChrysler is fully engaged in the development of these systems and will continue to be a leader in the field.

Lessons Learned

Over the decade or so of Chrysler's experience in the AFV arena, a number of lessons have been learned. If the knowledge gained from these programs is applied to hydrogen vehicle programs in the future, the resources of industry, government, and society can be most efficiently applied to move to a hydrogen economy.

One key lesson learned by Chrysler over the past decade is that any AFV program needs to be an integral part of a clearly articulated and broadly accepted national energy policy with explicit goals. The goals should delineate with some precision the problem that is to be tackled, whether it is CO_2 emissions, fossil fuel use, or dependence on imported petroleum. It must be recognized that the goals adopted will determine the relevant solution sets and may preclude some politically popular alternatives.

As part of the national energy policy, the goals and targets of the transportation sector must be set at achievable and measurable levels. These goals and targets should be measured against goals and targets for other segments of the economy, so that the most cost effective, least disruptive solutions can be pursued.

Next, any AFV program must rely on a total lifecycle analysis. All energy and resource inputs should be identified and quantified, not only for the vehicle when in use, but for the manufacture and ultimate disposal of the vehicle. The analysis must assess the energy and environmental consequences of the acquisition, processing, and distribution of the fuel, and for externalities associated with the program, such as resources required to secure the raw energy resource, local emissions benefits or detriments, societal impacts (including impacts on various socioeconomic groups), employment, and resource allocation.

An AFV program must be viewed as a system in order to succeed. Equal attention must be paid to the development and deployment of the vehicles, production and distribution of the fuel, and consideration for customer demands, whether they be range, behavioral changes, secondary markets, or safety, to name a few.

Furthermore, any AFV program must compete against the very mature spark ignited, internal combustion engine, fueled by petroleum derived gasoline in a world in which petroleum is abundant and inexpensive, cheaper in fact than any currently identified alternative fuel. Furthermore, the share of petroleum that is imported from politically unstable areas will continue to grow as long as the lowest cost sources are offshore. Government must recognize that it needs to play a role in establishing economic policies to "level the playing field." This may include providing new financial incentives or subsidizing both vehicles and fuels. As important as these incentives and subsidies are, they must be finite in duration. Any replacement for the gasoline powered spark ignited engine must eventually be cost competitive on a stand alone basis if it is to be successful.

Every major automaker must be willing to subsidize the development and early deployment phase of new technology introduction, but fiscal responsibility requires that there be some promise of a positive return in the future. In the specific examples discussed here, Chrysler lost money on CNG, M85, E85, and BEVs, with no prospect of ever recouping those investments. This is a pattern that no manufacturer can afford to continue indefinitely. The lessons learned from this experience, however, can reap enormous economic benefits as DaimlerChrysler prepares to commercialize clean diesel, hybrid electric, and hydrogen powered fuel cell electric vehicles.

CHAPTER 14

Understanding the Transition to New Fuels and Vehicles: Lessons Learned from Analysis and Experience of Alternative Fuel and Hybrid Vehicles

Paul Leiby and Jonathan Rubin

Since the energy crisis of 1973, the United States (U.S.) has explicitly sought to moderate the consumption and importation of oil. Initially the dominant concerns were energy conservation and energy security. During the 1980s and 1990s, the additional concerns of urban air quality and greenhouse gases took center stage. The transportation sector represents about 28 percent of total domestic energy use (Davis and Diegel, 2002, Table 2.1). Of the total amount of transportation energy used in the United States, the demand is overwhelmingly met by petroleum. As David Greene (1996) points out, the almost complete dependence of the transportation sector on petroleum persists today despite the market upheavals of the 1970s and early 1980s. In 1973, at the height of the Arab embargo, the U.S. transportation sector was 95.6 percent dependent on oil, about 1 percent less than today. Given geopolitical developments affecting global oil supply and renewed concern about the peaking of world oil production (Campbell and Laherrere, 1998), oil dependence is again a major driver of transportation energy policy.

The question remains: How should our transportation system transition itself to be more sustainable and secure? In this chapter we discuss the lessons about transitions that can be learned from past efforts to move to alternative transportation fuels, and what we can learn from the growing introduction of hybrid electric vehicle (HEV) technology. This is very important, since future policies and tax dollars are driven, in part, by expectations regarding the payoffs of emerging fuels and technologies such as hydrogen and fuel cells. Hydrogen fuels and fuel cell vehicles (FCVs) share many of the same attributes, promises, and challenges of other alternative fuel vehicles and electric drive vehicles, only perhaps magnified and compounded.

More particularly, this chapter presents the lessons learned from our modeling and analysis of the transition to alternative fuel vehicles (AFVs) and HEVs. What has worked, or could work, in a market based economy where economies of scale in vehicle production are significant and endogenous network effects are important? For the most part, the AFV technologies we examined were mature technologically, though not mature in the marketplace. For HEVs, we explicitly examined the role of technological learning to reduce production costs. It is our belief that explicitly evaluating transition paths, as opposed to static equilibrium outcomes, is very important and leads to compelling insights for a wide variety of economic and environmental questions. This of course implies that policies designed to guide markets must deal explicitly with transitions and not be solely concerned with future outcomes, which may in fact never occur.

The Goal of Reducing Petroleum Demand in Transportation

There are three ways to reduce the amount of petroleum used by cars and light trucks: lower the amount of driving per year, increase the average fuel efficiency of the vehicle fleet, or substitute alternative fuels for gasoline. The principal fuel efficiency measure in place in the United States is the corporate average fuel economy (CAFE) standards for new cars and light trucks (U.S. Energy Policy and Conservation Act of 1975). CAFE regulations specify minimum fleet average standards for fuel efficiency that vehicle manufacturers must meet. The Alternative Motor Fuels Act of 1988 (AMFA) provides credits for the manufacture of dedicated alternative and flexible fuel vehicles. The fuel substitution approach is advocated by Section 502(b) of the Energy Policy Act of 1992 (EPACT) and has been an important impetus for research, development, and demonstration (RD&D) projects and for fleet programs to help alternative fuels enter the U.S. transportation market. Other important alternative fuel incentives include gasoline composition and air quality requirements in the Clean Air Act Amendments of 1990 and large tax credits for ethanol.

EPACT also provides incentives to introduce AFVs and requires that the U.S. Department of Energy (DOE) estimate the technical and economic feasibility of producing sufficient alternative and replacement fuels to replace, on an energy equivalent basis, at least 10 percent of gasoline use by the year 2000, and at least 30 percent by the year 2010 (EPACT, 502[a], 502[b]). Petroleum is displaced by the use of neat alternative fuels as well as through the use of reformulated and oxygenated gasoline that contains natural gas, hydrogen, and alcohol and ether-oxygenates. Replacement fuels, loosely speaking, are those portions of gasoline that are not gasoline, such as oxygenates, and other nongasoline fuels that are not alternative fuels, such as gasohol, which is a blend of 10 percent ethanol and 90 percent gasoline. The recent strong interest in hydrogen-fueled vehicles reflects its promise for gasoline substitution, improved end-use fuel efficiency, and a set of diverse domestic supply options.

In 1996, the DOE published the results of their initial analysis of EPACT's goals, using the Alternative Fuels Trade Model (U.S. Department of Energy, 1996; Leiby, 1993). This study determined, among other things, that, "For the year 2000, 10 percent replacement of light-duty motor fuel use with alternative and replacement fuels is feasible and appears likely with existing practices and policies." The DOE report further states: "Displacing 30 percent of light-duty motor fuel use by 2010 appears feasible. However, this estimated feasibility is based upon a number of assumptions that may not be realized without additional alternative-fuel initiatives." These results are shown in Fig. 14-1.

In contrast to these 1993 estimations of long-run gasoline substitution, in 2001 replacement fuels contributed only 2.6 percent of total on-road motor fuel, on a gasoline gallon equivalent (GGE) basis. This is despite efforts by the DOE to promote alternative fuel use in government fleets and other markets and despite the millions of E85 flexible fuel vehicles (FFVs) on the road (Energy Information Administration [EIA], 1999, 2001). These FFVs can either run on straight gasoline or a blend of 85 percent ethanol and 15 percent gasoline. As described in detail below, it is also quite unlikely that the 30 percent displacement goal for the year 2010 will be met. This shortfall from the projections based on comparative-static analyses can largely be attributed to an incomplete understanding of the magnitude and importance of certain key dynamic or transitional impediments to alternative fuels. This earlier work provided critical foundations for our transitional analysis.

A Transitional Analysis

As recognized by the DOE's own analyses, past studies of alternative fuel and AFV penetration assumed either mature markets with large-scale vehicle production and the widespread availability of alternative fuels at retail stations, or immature markets and small scale production. Early studies of the AFV market can be grouped into those that are static, single year snapshots (National Research Council, 1990; U.S. Department of Energy, 1996)

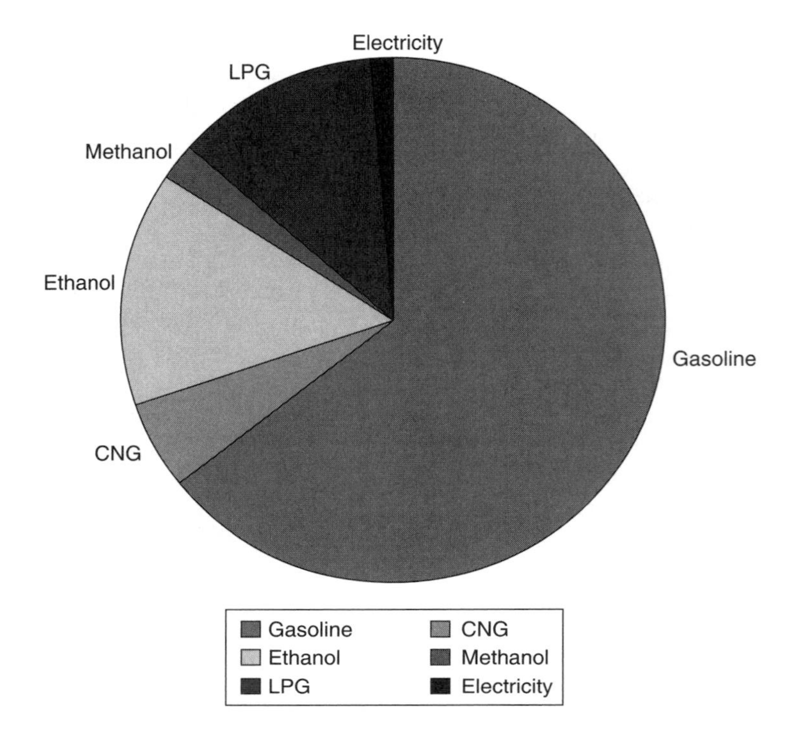

FIGURE 14-1. Results of static equilibrium analysis of alternative fuel penetration, year 2010 projected fuel shares. *Source:* Leiby, 1993.

and those that are multiyear analyses (Fulton, 1994; Rubin, 1994; Kazimi, 1997a,b), still with limited degrees of dynamic detail. Obviously, the static analyses are limited in that they cannot assess the feasibility or cost of a transition to the new long run equilibrium. Furthermore, in many cases, their conclusions, as well as those of most early dynamic models, reflect exogenous assumptions regarding fuel and vehicle prices or AFV penetration rates or both. Those results, in fact, can turn out to be misleading. This is because barriers to new fuels and technologies are real and, for the case of transportation technologies, economically important. Recognizing these issues, the DOE commissioned the creation of a transitional alternative fuel and vehicle model (TAFV).

The overall objective of the TAFV model is to assess the competitive market outcome over time, with and without possible new policy initiatives. Rather than taking fuel and vehicle prices and penetration rates as an input, they are determined from market conditions. Operationally, this is equivalent to maximizing consumer and producer surplus, or well-being, from transportation services provided by light duty vehicles, essentially cars and light trucks, and a variety of possible fuels. The TAFV model characterizes interactions among fuel providers, vehicle producers, fuel retailers, private vehicle purchases, and fleet vehicle programs. Each supply sector is represented by a single period cost function defined for each period, region, fuel, and vehicle type. Examples include vehicle production costs, fuel production or conversion costs, fuel retailing costs, raw material supply costs, and sharing or mix costs associated with vehicle and fuel choices. The cost functions summarize the way in which changing levels of activities, inputs, and outputs affect the costs for each supply module, and implicitly define the cost minimizing behavioral relationships among the model's variables.

Benefits in this model come from the satisfaction of final demand for transportation services as determined from projections of light duty vehicle fuel use, excluding diesel, for 1996 to 2010 given in the *Annual Energy Outlook* published by the U.S. Energy Information Administration (EIA). The total demand for light duty fuel is satisfied by the use of existing vehicles and the purchase and use of new vehicles. The use of older vehicles is limited by the stock of each vehicle type given a fixed, age-adjusted use profile. Each year, some vehicles are scrapped, and, to the extent that existing vehicle stocks are insufficient to satisfy the demand for transportation services, a mix of new vehicles is purchased. New vehicles are chosen according to a nested multinomial logit (NMNL) choice formulation, whose parameters come from David Greene's work (Greene, 1994). Vehicle choice is based on upfront vehicle capital costs, nonprice vehicle attributes, and expected lifetime nested fuel choice costs. Fuel choices must be made for the vehicles that are dual or flexibly fueled. Since vehicle and fuel choice is endogenous, it is important to specify which fuel and vehicle characteristics are considered in the fuel and vehicle choice submodules and which characteristics are endogenously determined. A more detailed presentation of assumptions and data sources can be found in Leiby and Rubin (1997, 2000, 2001).

From preliminary analysis and discussions with experts, we identified key areas that could strongly affect the transition to alternative fuels and vehicles. These include the costs to consumers of limited fuel infrastructure and retail availability of alternative fuels; scale economies for vehicle production and fuel retailing; the implications for consumer behavior of initially limited AFV model choice and diversity; the prospect for technological improvement and cost reduction through learning-by-doing; and any costs to consumers from being unfamiliar with a new technology. Because of their potential importance, all these transitional barriers, except for those related to consumer unfamiliarity, have been explicitly modeled. We did not model the costs of consumer unfamiliarity for new technologies since we had little information to make realistic parameter estimates. As our results below suggest, omitting this additional possible cost is unlikely to have altered any of our qualitative results since the AFV market has a difficult time getting started given the transitional barriers that we do include.

Effective Costs of Limited Retail Fuel Availability

Most alternative fuels are currently available at only very few retail stations. First principles and evidence from surveys of diesel car buyers (Sperling and Kurani, 1987) suggest that fuel availabilities, or station shares below 10 percent, can impose large effective costs on consumers. There is, however, little empirical evidence as to the possible size of these costs. Our approach is to use work by Greene (1998), who asked the following question in two national surveys: "Suppose your car could use gasoline or a new fuel that worked just as well as gasoline. If the new fuel costs $0.25, $0.10, or $0.05 less per gallon but was sold at just one in 50, 20, or 5 stations, what percent of the time would you buy this new fuel?"

Greene used a variety of functional forms to estimate a random utility, binomial logit choice model. He found it hard to discriminate among functional forms based solely on the degree of fit to data and noted that we are particularly uncertain about the effective costs of fuel availability at very low levels, say less than 5 percent station share. In addition to issues of fit, we have chosen to use Greene's exponential functional form because our intuition tells us that at 50 percent fuel availability, meaning every other gas station, the cost penalty ought to be small. For the exponential functional form, the cost penalty at 50 percent availability is $0.02 per gallon. At less than 0.1 percent fuel availability, the cost per gallon is uncertain, but using the exponential functional form is conservatively estimated at $0.35 (see Fig. 14-2).

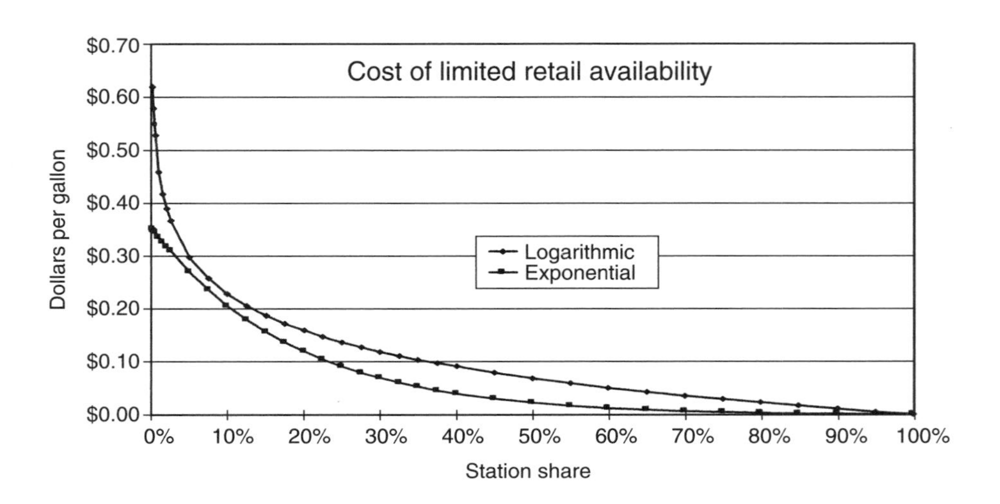

FIGURE 14-2. Effective cost to consumers versus fuel availability.

Vehicle Manufacturers' Costs per Model and Economies of Scale

The TAFV model is designed to estimate the costs of vehicle production for the following alternative fuels: liquefied petroleum gas (LPG), compressed natural gas (CNG), methanol, ethanol, and electricity. The vehicles are either dedicated to a particular fuel type or are capable of using both gasoline and the respective alternative fuel. The one exception is electricity, which is a direct fuel input only to dedicated electric vehicles. AFV costs are calculated from engineering–economic estimates of the incremental cost of each AFV fuel technology compared to conventional vehicle technology (Energy and Environmental Analysis, 1995). The AFV technologies that we model, except for electric vehicles and HEVs, are assumed to be mature. This means that for a given production scale further production experience will not reduce per unit vehicle production costs at a rate significantly faster than it would for conventional vehicles.

Substantial economies of scale in vehicle production exist in the automotive industry. That is, there are sharp reductions in per unit vehicle cost with larger volume production of any given vehicle model. Significant scale economies for a given vehicle model occur in around 10,000 vehicles produced per year, with full economies achieved at 100,000 vehicles per year. Therefore, we modeled per unit vehicle production costs as a declining function of average vehicle plant size in each year. The volume of production in any given year is constrained by the level of cumulative investment in production capacity less capacity decay. This means that vehicle prices and manufacturing capacity are endogenous variables. This has the advantage of admitting the positive feedback effects from policies that encourage the early adoption, and hence larger scale production, of AFVs.

Accounting for Vehicle Model Diversity and the Effective Cost of Limited Diversity

Consumers contemplating buying a new gasoline fueled car are offered a wide variety of vehicle makes and models with a huge number of features to choose among. The marketwide attractiveness of an alternative fuel or hybrid electric technology will depend on the diversity of vehicle models for which it is made available. Offering, for example, methanol fuel technology on only a single model will put methanol vehicles at a disadvantage compared to gasoline vehicles, all else being equal. At the same time, offering methanol capability on several different models is expensive because it lowers plant scale for any overall level of production. Thus, during the transition period for a new vehicle technology, there is an inherent tension between providing model diversity and achieving economies of scale.

Rather than predetermining the number of makes and models offered with alternative fuel capability, the TAFV model balances the additional production costs against the additional consumer satisfaction. This is accomplished by defining a variable that represents the number of makes and models for each vehicle and fuel type produced. On the vehicle production side, we divide the total industry production capacity for each vehicle and fuel type by this diversity variable to get the average production scale for that vehicle type; on the consumer side we incorporate the diversity variable into our multinomial choice framework.

The value of diversity to consumers depends on the order in which vehicle manufacturers introduce a new fuel or vehicle technology to their existing model lines. This is because different models have market penetrations that vary from a few thousand per year for specialty cars to well over 100,000 for some popular cars and pickup trucks. If alternative fuel capability is introduced randomly on different vehicle models, then we estimate the cost to be initially $2080 per vehicle. If manufacturers add the alternative fuel technology to the most popular model line first, then the effective cost to consumers of limited model diversity is lower, dropping to an estimated $727 per vehicle. When AFV offerings ultimately have the same richness of models as gasoline vehicles, then the limited diversity cost is nothing. In the simulation model we assume that the alternative fuel technology is offered on the most popular model first. Thus, the significance of the initially limited model diversity for new vehicle technology depends on the strategy that manufacturers adopt when introducing the technology. If the technology is introduced on a new, unfamiliar, and made-to-purpose vehicle, such as Honda initially did with the Insight HEV, then the deterrent effect of limited model diversity is far greater than if the manufacturer takes a chance by introducing the technology on its most popular vehicle, such as Honda subsequently did with the Civic Hybrid.

Technological Cost Reduction and Learning-by-Doing

Learning-by-doing (LBD) is the process by which the costs of new technologies decline as a function of cumulative experience. This phenomenon is observed in various industrial situations where it is described as a learning curve, progress function, or experience curve. It is important to distinguish LBD, scale economies, and learning by technological progress, which is closely related to R&D spending.

The theory of LBD was first exposited by Arrow (1968). Empirical studies using historical data suggest learning rates in the range of 5 to 20 percent per doubling of experience (e.g., Lieberman, 1984; International Energy Agency, 2000; McDonald and Schrattenholzer, 2001). At the same time, however, these rates must be used with great caution. This is because, as McDonald and Schrattenholzer note, the empirical literature varies in its methodologies and data sources by which learning rates are calculated. For example, the literature does not always disaggregate learning from the effects of scale or R&D expenditures. Sometimes the dependent variable is price, rather than cost, and price is influenced by supply and demand factors not related to learning. Finally, the period chosen for the empirical analysis can also affect the calculated learning rate. Notwithstanding these limitations, LBD, as documented in the empirical literature, appears to be an important component of technology change and cost reduction. The existence of substantial learning may also be important for determining good public policies designed to spur new technologies.

A concept related to LBD is learning from R&D. The Partnership for a New Generation of Vehicles and FreedomCAR programs are classic examples of this research based approach to advanced automotive design. Public policies to encourage new technologies can encourage both R&D and LBD. However, the prospect of learning from R&D can have significantly different policy implications than LBD. As Goulder and Mathai (2000) show in the context of climate change, if knowledge is gained primarily through R&D, then it may be justifiable to shift abatement to the future and to act later. If cost reductions are also gained via LBD, then the impact of learning and technological change on the timing of abatement efforts is ambiguous. The same reasoning may be applied to policies for promoting new vehicle technologies. Depending on the particular technologies and assumptions, it may be optimal to act sooner and implement technologies, to learn and thereby lower future costs. If the endogenous LBD rate is sufficiently rapid, LBD proponents argue that forcing a sharp divergence from existing technologies by performance mandates could induce otherwise uneconomic technologies to become economically viable.

For the TAFV modeling results described here, we assumed that the prospects for nonhydrogen AFV cost reduction through LBD were no greater than those for conventional vehicles, based on the views expressed by those who provided the AFV cost estimates (Energy and Environmental Analysis [EEA], Inc.). This means that the incremental costs for AFVs relative to conventional vehicles did not improve through LBD. For HEVs, on the other hand, we assumed that given the comparative newness and complexity of the technology, a moderate rate of LBD could be gained at a rate of a 10 percent reduction in cost per doubling of cumulative production experience. The net effect implied is that by the time HEV production reaches a significant share of total new vehicles produced, HEV incremental costs for large plants could decline to about one-half of their initial level.

Lessons Learned about the Transition to New Fuels and Vehicles

Despite the range of federal, state, and local initiatives to promote alternative fuel use since the 1980s, there has been little progress in developing alternative fuel infrastructure, advancing the alternative fuel transition, or achieving alternative vehicle sales and fuel sales. Apart from the mandated sales of AFVs to some fleets, the principal exception has been the sale of over 4 million ethanol FFVs. The sale of FFVs, mostly to private consumers, is completely understandable, and was predicted by the TAFV model, as the most profitable response of domestic vehicle manufacturers to the federal provision of CAFE credits in the AFMA (Rubin and Leiby, 2000). However, as was also anticipated, an associated ethanol supply infrastructure has not developed, since these vehicles use little or no ethanol and their FFV property is virtually irrelevant to most owners, except in selected states with heavy ethanol subsidies. These historical facts alone provide useful lessons about the prospective transition to hydrogen fueled vehicles. The transitional analyses done with the TAFV model help to explain these outcomes and allow investigation of possible future outcomes under different policies and incentives, as well as different market conditions and oil price regimes.

In analyzing the transition to alternative fuels other than hydrogen—such as ethanol, methanol, CNG, LPG, and electricity—analyses with the TAFV model led to some important conclusions that bear on the proposed hydrogen transition. We find that the transition matters a lot. Furthermore, we can identify some of the most important barriers. For AFVs, the most important barriers seem to be limited fuel availability and vehicle scale economies. For HEVs, incremental vehicle costs are large. As a result, vehicle scale economies matter, but scale cost reductions are more easily attained by the use of widely shared components—such as batteries, motors, and controllers—across multiple vehicle platforms. Similar gains should be possible for FCVs. For HEVs, the dominant transitional factor is the uncertain prospect for LBD.

Both the real-world experience with federal and state fleet AFV programs over the last decade and modeling analyses of proposed expanded private and local fleet mandates indicate that forcing vehicle technology adoption by a few doesn't work unless the technology has private appeal. This does not bode well for the strategy of mandated or heavily subsidized niche introduction.

Transitions Matter a Lot

The barriers to new fuels and technologies are real and economically important. Some barriers are transitional, and some barriers will endure so long as overall market conditions—including oil prices and environmental policies—do not change fundamentally. We find that static equilibrium analysis of the prospects for new vehicle technologies can be misleading.

Modeling experiments confirm that transitional barriers, and the particular transition paths pursued, matter a lot for the technology's ultimate market success. Using similar technology and market assumptions as the original U.S. EIA (1993) study on alternative fuel vehicles, we reestimated the likely market penetration of alternative fuels by 2010 using a dynamic analysis and incorporating transitional factors. We incorporated the transitional factors discussed above, including vehicle and production scale economies, consumer costs of low retail fuel availability, consumer costs from limited AFV model choice, and slow turnover of the on-road vehicle fleet. We did not assume any substantial LBD for the AFV technologies considered, given the maturity of the technologies. LBD is more likely for electric drive vehicles.

Figure 14-3 shows the projected 2010 fuel shares, accounting for transitional barriers. These data suggest that the market share of alternative fuels may be small, absent significant oil price change or large and new incentives. Compare with Fig. 14-1, which shows the results of the same analysis but omitting transitional barriers.

The new results, accounting for the transition and potential barriers to alternative fuels, are strikingly different from the long-run equilibrium analyses. We found that transitional barriers will prevent any significant alternative fuel use by 2010 without sustained, expensive market interventions. Moreover, we estimate that these market barriers are approximately equivalent to a cost of $1.00 per gallon today, and persisting at a level of $0.50 per gallon by 2010. Finally, absent major new government policies to promote alternative fuels or reduce greenhouse gases, it is unlikely that the

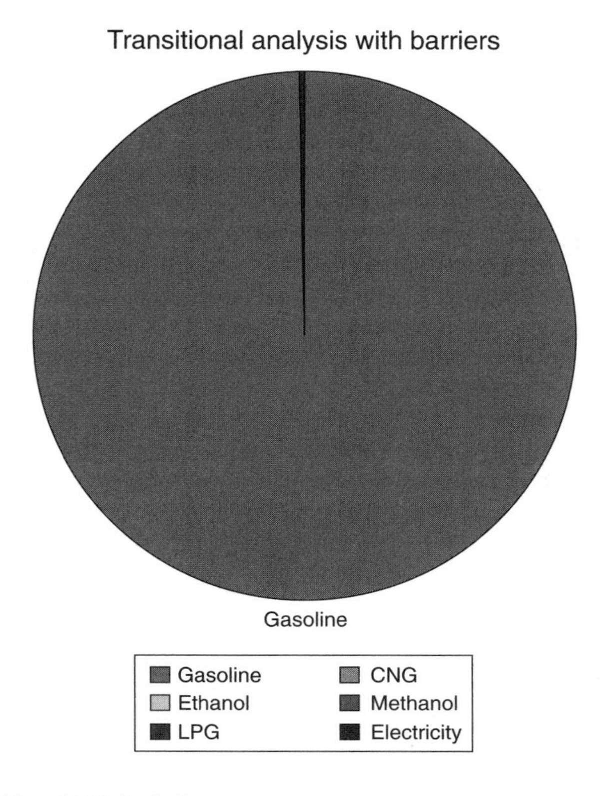

FIGURE 14-3. Year 2010 fuel shares.

United States will achieve or even approach EPACT's 2010 displacement goals (Leiby and Rubin, 2001). These transitional barriers are likely to have similar implications for prospects of hydrogen ICE vehicles and for FCVs.

Coordination, Choice, Scale, Learning

There is a conflict between diversity of choice and cost savings resulting from economies of scale in vehicle production and network economies for fuel provision. This means that there is a trade-off between offering the consumer many vehicle and fuel choices and achieving low cost transportation services. For AFVs, initial vehicle costs at low scales are 10 to 50 percent higher than their full scale costs, and achieving scale economies quickly is essential for profitability. Taking a broader view, it means that at the level of the market selection of vehicle and fuel technology, there is a major tension between the merits of producing different AFV technologies for different circumstances and for different consumer regions or market segments, and the substantial scale and network economies associated with producing and fueling a single vehicle-fuel technology.

This tension proves to be more important than most would anticipate, particularly when we account for transitional barriers. In the long run comparative static analyses of 1993, it was estimated that as many as five different AFV technologies would compete and contribute simultaneously to displace significant fractions of gasoline. The modeling analysis that included transitional issues repeatedly found that it was difficult for any AFV to gain market share. Moreover, only one or two alternative fuels at most could simultaneously exist in the market when gasoline was displaced (see Fig. 14-4). This point also has significance for the problematic enterprise of "picking winners," in other words choosing particular technologies to promote. If only one or two new technologies can survive in the challenging network of vehicles and fuel, the cost of choosing and promoting what might be the wrong one is very high.

Forcing a Relatively Small Market Segment to Buy AFVs Will Not Work

The DOE has the authority under EPACT to require private fleets and those of state and local (P&L) governments to purchase AFVs, expanding the fleet AFV sales mandate to about 2 percent of total vehicle sales. We estimate the outcome of imposing the P&L fleet mandate to be that nonfleet vehicle owners would be induced to purchase an additional 2.9 percent of AFVs by 2010. Thus, under this rule a total of 4.9 percent of new vehicle sales would be AFVs by 2010. Thus, the P&L fleet rule could help reduce transitional barriers by lowering the cost of AFVs. Unfortunately, the vehicles chosen are mainly alcohol FFVs, which use very little alternative fuel given their high cost. Furthermore, the fleet and private demand for AFVs encouraged

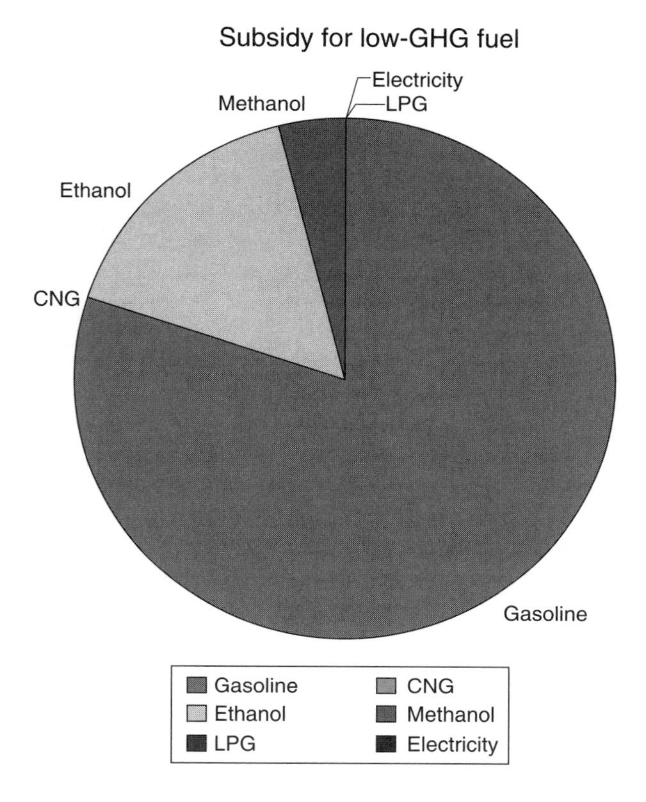

FIGURE 14-4. Year 2010 fuel shares for case of large subsidy for low greenhouse gas fuels.

by the EPACT mandates crowds out the induced value of the CAFE credits. That is, the AFV demand induced by the fleet mandate far exceeds the number of AFVs eligible for the CAFE credit, and the credit value falls to zero. As with the base case, fuel price sensitivity analysis shows these results to be quite robust.

More fundamentally, this type of mandate does not work because you need a product with superior private value for mandates or subsidies to effectively induce voluntary adoption of the technology. This policy does not provide additional value and does not address the issue of fuel retail availability. Again, absent compelling advantages in private value, sustained and significant policy intervention is required. Moreover, the P&L mandate would be quite costly. As can be seen in Fig. 14-5, the cost of an expanded fleet AFV mandate is as much as \$1.40 per gallon gasoline-equivalent (GGE) displaced under current oil prices. In this figure, cost is measured as net social surplus loss per GGE of gasoline displaced, excluding externalities.

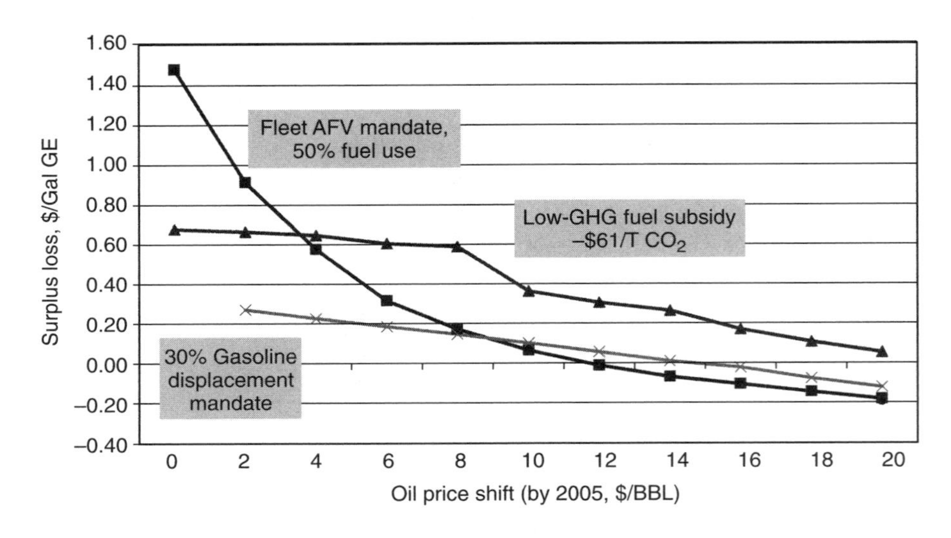

FIGURE 14-5. The cost of AFV policy tools versus oil price levels.

Both the real-world experience with federal and state fleet AFV programs over the last decade and modeling analyses of proposed expanded private and local fleet mandates indicate that forcing vehicle technology adoption by a few doesn't work unless the technology has private appeal. This does not bode well for the strategy of niche introduction given the experience to date and the model implication that it is hard for niche vehicles to move out of the niche. Moreover, statements by fleet managers and vehicle manufacturers emphasize that fleet owners are very sensitive to the lifecycle economics of their vehicles, including their residual, or resale, value in the general market after a few years of service. Commercial and government fleets, which often rely on quite conventional passenger vehicles, prove to be a demanding market for new vehicle types, rather than a natural or easy market.

Importance of Oil Prices as a Long-Run Barrier

The comparatively low price of oil projected by many government agencies (e.g., Energy Information Administration, 2003), both relative to the history of the last few decades and relative to the current and projected price of alternative fuels, presents a significant and sustained barrier to new transportation fuels. Overall, without compelling private advantages, large and sustained policy intervention is required to attain substantial market share for new fuel and vehicle technologies, particularly AFVs or full HEVs operating at 300 volts (V) or higher. Modeling analyses also indicate that a temporary oil market shock, even when coupled with moderate subsidies or vehicle sales mandates, cannot induce significant AFV market share so long as consumers anticipate that the oil price increase will fade away after 2 years.

Without eliminating this long run cost barrier, transitional policies can be ineffective and costly. Figure 14-5 shows how the cost of gasoline displacement by AFVs is high at current prices, but would decline if the relative cost of oil were to increase. Alternatively, under sustained higher oil prices, transitional policy can work. In these circumstances, both price and nonprice incentives can be effective, as shown in Fig. 14-6. Figure 14-6 shows the estimated share of gasoline displaced in 2010 by (nonhydrogen) alternative fuels for a range of higher oil price levels. The horizontal axis, labeled "Oil price shift," refers to the increase of oil prices above the EIA base path, achieved gradually over the next 5 years and sustained thereafter. Similar market share results hold if sustained higher costs are imposed for carbon emission at levels that provide an equivalent price differential for alternative fuels.

Some strategies that are quite costly and ineffective under current market conditions become quite efficient at inducing new vehicle technologies when coupled with a permanent oil price increase or a major change in environmental policy. Early vehicle and fuel use mandates in selected fleets, and possibly other niche applications, are costly under current market conditions, but they appear to set the groundwork for rapid technology gains and infrastructure development if market conditions, such as the cost of oil or carbon emissions, change radically and enduringly. Thus, these high

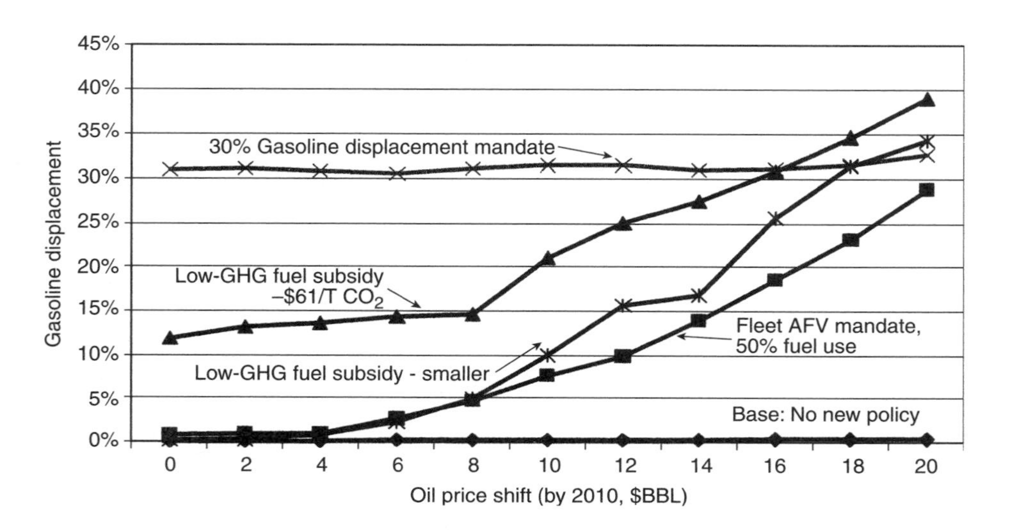

FIGURE 14-6. Effectiveness of policy tools at different oil price levels (2010 alternative fuel share vs. oil price shift).

cost programs have option value in the face of energy market and environmental uncertainty.

Simulations also suggest that a deployment mandate or niche application can be an effective strategy if one anticipates substantial technological LBD. This is relevant for the hydrogen problem.

The Ability of Conventional Technology to Adapt Can Forestall the Transition

The U.S. experience since the 1988 AFMA and the 1990 Clean Air Act amendments has shown how hard the transition to new motor fuels can be. It has also shown that established, conventional technologies and industries can and will respond to the challenge of a substitute vehicle or fuel. The pressure for clean, alternative fuels was answered by the development of reformulated gasoline. Some of the newly formulated conventional fuels embodied alternative fuel components and blends, providing a path for a measure of renewable and alternative fuel use in existing vehicles without the need for wholly new fuel distribution infrastructures or markedly new vehicle designs.

Vehicle manufacturers also demonstrated the ability to produce increasingly clean, gasoline burning engines, sharply reducing the air quality motivation for alternative fuel vehicles. Manufacturers ultimately produced near-zero emission vehicles (ZEV) that challenged the need for the ZEV mandates in some states and diminished the pressure to produce electric vehicles. This demonstrated ability of the fuel and vehicle industries to respond by adapting conventional systems has not been included in the TAFV modeling analyses, which considered a range of vehicles but took vehicle designs as static. However, provision for this natural market response should be made in the planning and analysis of potential future transitions.

Technology Change Important, Yet a Poorly Understood Issue for the Transition to Hydrogen

Learning curves are widely believed to be important and significant for many new technologies. But representing learning is tricky, and its magnitude and timing is not well understood. In addition, if one accounts for cost reductions through LBD when assessing policy for a technology transition, a balanced approach would also consider the potential for two countervailing effects. The first is the learning or adaptability of conventional technology. The second is that the potential for forgetting, the flip-side of learning, can negatively affect future outcomes.

The choice of an index of experience to use for LBD has long been a topic of research. In TAFV, there are several ways to model cumulative experience: as cumulative new production capacity or new vehicles sold, or as total installed production capacity or the total stock of vehicles on the road. Given that production capacity and vehicles are scrapped over time, the latter two methods allow for forgetting as well as learning. Globerson and Levin (1987) have argued that we should incorporate both learning and forgetting into institutional environments. Benkard (2000), drawing on data from the aircraft industry, showed that in certain industries there is evidence of organizational forgetting, saying that "...production experience actually depreciates over time, and knowledge gained from building one product doesn't necessarily spill over to the next generation." Benkard found that a model that includes depreciation of experience accounts for the data much better than the traditional learning model.

We model LBD in terms of total installed vehicle production capacity. This assumption allows both accumulation and depreciation of experience, and does affect our results. Short lived vehicle subsidies, if they are insufficiently large or long in duration, may be unable to induce a sustainable HEV production sector. Such temporary subsidies may only temporarily lower vehicle production costs from learning.

Without subsidies we project no substantial penetration by HEVs, based on their prospective fuel efficiency gains and costs alone. This may reflect an overly pessimistic assessment of the costs of hybrid electric technologies. If so, our results should be interpreted to hold with lower levels of subsidies than indicated. Hybrid subsidies on the order of $2000 per vehicle can induce substantial hybrid penetration and gasoline demand displacement under EIA's 2001 oil price projections (see Fig. 14-7). This result is quantitatively different from the result achieved for AFVs. The efficacy of

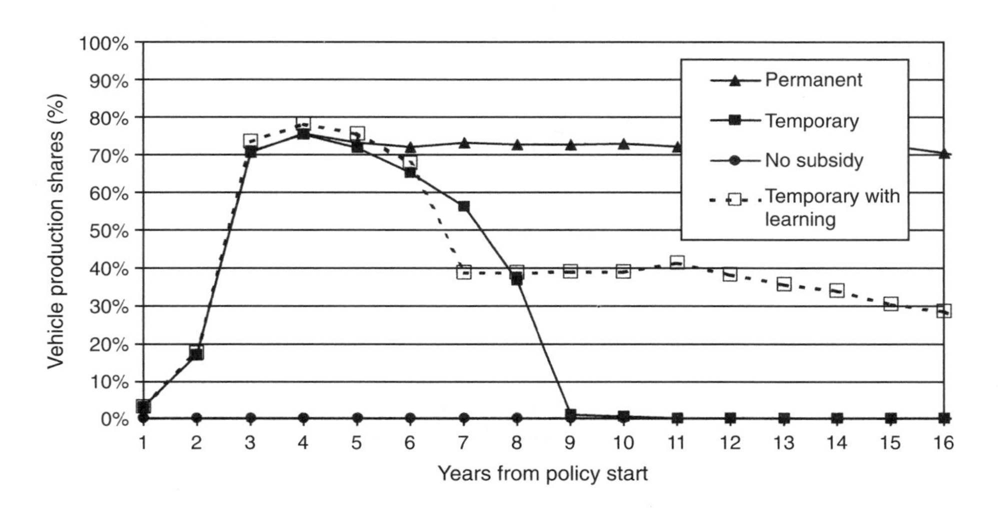

FIGURE 14-7. HEV production share with permanent and temporary ($2400) subsidy, and with temporary subsidy given learning by doing.

HEV subsidies is much greater than for AFVs because of the latter's huge fuel infrastructural needs. The HEVs sold are primarily of the "mild" hybrid type with 42-V systems that achieve modest fuel efficiency gains of 15 percent with greater cost effectiveness than more extensive hybrid designs operating at much higher voltages. Temporary HEV subsidies are effective at inducing HEV penetration but do not have long term effects once they are removed unless there are cost reductions due to LBD. With LBD we found that a high enough temporary subsidy can be effective in assisting HEVs to overcome transitional barriers and become self-sustaining in a competitive market.

Clearly LBD is an important factor. Its importance for costs has been validated in the literature and confirmed in empirical studies of many industries. Its importance for new vehicle policy was validated again here. But, while important, learning is very tricky to represent. For this reason, in order to better understand the prospects for transitions to new vehicle technologies, including AFVs, hybrids, and fuel cell vehicles, further work is needed to refine the estimates of technology costs and learning rates.

Lessons and Challenges for Fuel Cell Vehicles: Is Hydrogen Different?

One might ask whether the challenge of the transition to hydrogen vehicles is different from that of other alternative fuel or hybrid vehicles. In some ways the answer is yes, particularly since there are substantial prospects for nontransportation applications of hydrogen and fuel cells, and the opportunity for some synergistic development and efficiency gains between the sectors. Also, FCVs involve a host of new supply, distribution, storage, and vehicle technologies, some of which still have the possibility of marked improvement and cost reduction.

On the other hand, one could argue that hydrogen FCVs are much the same as AFVs, facing all of the same challenges, only more so. FCVs face the same monetary and business plan challenges of new technologies, like HEVs, and the same fuel system challenges of AFVs, exacerbated by the particular properties of hydrogen gas. Finally, they face the same doubts about political sustainability.

One clear lesson from our analysis of AFVs and HEVs is that an assessment of hydrogen's potential as a fuel for FCVs must consider transitional barriers. A simple, mature scale cost assessment will be misleading. Although there has been considerable speculation, and some scenario development, there has been little rigorous analysis of alternative ways the transition to hydrogen could take place. At present, no modeling system exists that can integrate the development of infrastructure, motor vehicle production, and the evolution of demand, as well as the necessary policy context, over time and geographically within the United States. Potentially important interactions with other energy sectors such as electricity generation and the spatial and geographic detail in the location of production, delivery, and demand are very important issues that are clearly essential for a fully satisfactory representation of the transition to hydrogen. As with hybrid vehicles, modeling endogenous LBD and exogenous R&D will be important.

On the hydrogen production side, the existence of multiple potential supply pathways and technologies is important to consider in the context of economics. One of the conclusions from the AFV transition analysis is that cost savings attainable with specialization of vehicle production and fuel infrastructure outweigh the estimated benefits of diversity of fuel and vehicle types. This also may be true for hydrogen production pathways. On the other hand, the diversity of potential feedstocks across the United States may allow different regions to specialize in particular supply and demand pathways. One possible issue for consideration is whether early movement to a small scale on-site production system would advance or impede larger scale hydrogen use over the long term.

Given the magnitude of the costs involved in transitioning to hydrogen FCVs and the possible links to the electricity generating sector, transitional modeling may require a general equilibrium framework as is currently common in models evaluating the regional and global effects of GHG reduction policies (Manne *et al.*, 1995). Such a framework would account not only for the interactions among many economic activities in multiple sectors, but also for the aggregate effect of large scale investments in the hydrogen system on tax flows and on the availability of capital for investment in other important competing economic activities, such as consumer goods production, health care, or environmental management.

On the demand side, an important transitional modeling issue involves how to account for the additional amenity value of fuel cell vehicles, including quiet operation, standby electricity generation power, and "green" value. At the same time, our AFV modeling work has shown the importance of recognizing the "disamenity" of limited vehicle model choice and limited fuel retail availability. A good characterization of these issues would be able to indicate which fuel pathways and vehicle technologies provide the greatest value to consumers.

Finally, given the magnitude of costs involved in a transition to hydrogen FCVs, an important consideration not explicitly addressed in the TAFV model is likely to be the role of political commitment. Political commitment, or perceived lack thereof, will have a direct economic cost in terms of increasing uncertainty, raising the costs for fuel and vehicle suppliers who make investments in equipment that will require use for many years to justify.

Conclusions

Overall, the market barriers to significant alternative fuel and vehicle use are substantial. We find that in the absence of any new and substantial policy initiatives, it may be difficult for the alternative vehicle and fuel markets to get started. Hydrogen FCVs face the same transitional problems and barriers that are shared by AFVs, only more so.

Our results lead us to several observations. First, in a market economy where vehicle manufacturers, fuel suppliers, and consumers all make independent decisions, the efficacy of government policies to reduce the dependence of the U.S. transportation sector on petroleum is highly dependent on the world price of petroleum. Second, the penetration of alternative fuels and AFVs depends on the fuel retail infrastructure, the ability of AFVs to achieve scale economies, and other transitional barriers. Third, governmental policies, if sufficiently large, can effectively reduce these barriers and can allow alternative fuels to compete in the marketplace with gasoline. However, given the current and expected low price of petroleum in the world today, doing so would be costly.

Given the worldwide nature of energy markets, there is a need to assess the competition from conventional vehicles and fuels, and consider the response. Since the marginal production cost of oil is $2 to $5 per barrel in the Middle East, and transportation costs at $1 to $2 per barrel are also low, there is a persistent threat that the economics of alternative fuels, including hydrogen, could be undercut by oil supplier actions. If significant quantities of gasoline and diesel are displaced, their prices can and will most likely decline substantially. This is a natural consequence of the multiproduct nature of the refining process and limits to the flexibility of product mix during refining operations. This may be offset, however, by rising demand for gasoline and petroleum in China and other developing nations.

Lest we appear too definitively negative about the transition toward AFVs and FCVs, we would like to make a few more positive notes. First, there is obviously a large potential role for advances in technology to bring down the costs and increase the benefits of hydrogen fuel cell vehicles. In addition, we have tried to model the existing vehicle market within the current and possible future regulatory context. Growing energy security or environmental concerns could motivate sufficiently strong policies to achieve the transition. For example, were the United States to ratify the Kyoto protocol and require reductions in greenhouse gases from the transportation sector on the order of 20 percent by 2010, then the whole price regime for transportation would be fundamentally altered, potentially allowing AFVs and FCVs to be much more competitive.

Efforts to pick winning alternative fuel technologies have fared poorly, as demonstrated with California's experience with methanol. Furthermore, the network nature of the motor fuel and vehicle markets, large economies of scale, and the prospect for learning through experience indicate that the least cost and natural market outcome is for a single or few closely related vehicle and fuel technologies to dominate. This suggests that policies for transitions should address targets, not technologies, since unexpected technologies may be superior to meet the goals, and second best technologies may have little long term market prospect.

Acknowledgments

We thank the following individuals for their contributions to this work: David Bowman, David Greene, Barry McNutt, David Rodgers, K. G. Duleep, Marie Walsh, and Thomas White. This work was supported in part by the DOE, Office of Energy Efficiency. The opinions and conclusions expressed in this chapter are solely those of the authors and do not represent those of the DOE or the authors' affiliated institutions. Additional information on the transitional analysis of alternative and hybrid vehicles can be found at: http://pzl1.ed.ornl.gov/altfuels.htm.

References

Arrow, K. J. 1968. "Optimal Capital Policy with Irreversible Investment." In J. N. Wolfe (ed.), *Value Capital and Growth: Papers in Honor of Sir John Hicks*. Edinburgh: Edinburgh University Press.

Benkard, C. Lanier. 2000. Learning and Forgetting: The Dynamics of Aircraft Production. *American Economic Review*, 90(4): 1034–1054.

Campbell, C. J., and J. Laherrere. (1998). The End of Cheap Oil. *Scientific American*, 278(3): 78–83.

Davis, Stacy C. and Susan W. Diegel. 2002. *Transportation Energy Data Book 22*. Oak Ridge, TN: Center for Transportation Analysis, Energy Division, Oak Ridge National Laboratory, U.S. Department of Energy.

Energy and Environmental Analysis. 1995. "Specification of a Vehicle Supply Model for TAFVM." Washington, DC: Energy Information Administration, U.S. Department of Energy, 1994.

Energy Information Administration (EIA). 1999. "Alternatives to Traditional Transportation Fuels: An Overview1993." Washington, DC: Energy Information Administration, U.S. Department of Energy.

Energy Information Administration. 2001. "Alternatives to Traditional Transportation Fuels 2001." Washington, DC: Energy Information Administration, U.S. Department of Energy.

Fulton, Lewis M. 1994. "Alternative-Fuel Vehicles and the Energy Policy Act: A Case Study in Technology Policy." Ph.D. Dissertation. University of Pennsylvania.

Globerson, S. and N. Levin. 1987. Incorporating Forgetting into Learning Curves. *International Journal of Operations and Production* 7: 80–94.

Goulder, L. H., and K. Mathai. 2000. Optimal CO_2 Abatement in the Presence of Induced Technological Change. *Journal of Environmental Economics and Management* 39: 1–38.

Greene, David L. 1994. "Alternative Vehicle and Fuel Choice Model." *ORNL/TM-12738* (October).

Greene, David L. 1996.*Transportation and Energy*. Eno Transportation Foundation, Inc.

Greene, D. L. 1998 Survey Evidence on the Importance of Fuel Availability to the Choice of Alternative Fuels and Vehicles. *Energy Studies Review* 8(3): 215–231.

International Energy Agency. 2000. *Experience Curves for Energy Technology Policy*. Paris: OECD.

Kazimi, Camilla. 1997a. Evaluating the Environmental Impact of Alternative-Fuel Vehicles. *Journal of Environmental Economics and Management* 33: 163–185.

Kazimi, Camilla. 1997b. Valuing Alternative-Fuel Vehicles in Southern California. *American Economic Review* 87(2): 265–271.

Leiby, Paul N. 1993. "A Methodology for Assessing the Market Benefits of Alternative Motor Fuels." Oak Ridge, TN: Oak Ridge National Laboratory, ORNL-6771, September.

Leiby, Paul N. and J. Rubin. 1997. The Transitional Alternative Fuels and Vehicles Model. *Transportation Research Record* 1587: 10–18.

Leiby, Paul N. and Jonathan Rubin. 1999. "Sustainable Transportation: Analyzing the Transition to Alternative Fuel Vehicles" *Transportation Research Board Circular, Transportation, Energy, and Environment* No. 492 (August).

Leiby, Paul N. and Jonathan Rubin. 2000. "The Alternative Fuel Transition: Results from the TAFV Model of Alternative Fuel Use in Light-Duty Vehicles 1996–2010." *Final Report - TAFV Version 1*, ORNL/TM-2000/168 (May 4).

Leiby, Paul N. and Jonathan Rubin. 2001. Effectiveness and Efficiency of Policies to Promote Alternative Fuel Vehicles. *Transportation Research Record*, 1750: 84–91. Transportation Research Board, National Research Council.

Lieberman, M. 1984. The Learning Curve and Pricing in the Chemical Processing Industries. *Rand Journal of Economics* 15: 213–228.

Manne, A., R. Mendelsohn, and R. G. Richels. 1995. MERGE: A Model for Evaluating Regional and Global Effects of GHG Reduction Policies. *Energy Policy* 23: 17.

McDonald, A., and L. Schrattenholzer. 2001. Learning Rates from Energy Technologies. *Energy Policy* 29: 255–261.

National Research Council. 1990. *Fuels to Drive Our Future*. Washington, DC: National Academy Press.

Rubin, Jonathan. 1994. Fuel Emission Standards and the Cost Effective Use of Alternative Fuels in California. *Transportation Research Record: 1444*, Transportation Research Board, National Research Council.

Rubin, Jonathan and Paul Leiby. 2000. An Analysis of Alternative Fuel Credit Provisions of U.S. Automotive Fuel Economy Standards. *Energy Policy* 28(9): 589–602.

Sperling, Daniel and Kenneth Kurani. 1987. "Refueling and the Vehicle Purchase Decision: The Diesel Car Case." *SAE* Technical Paper Series No. 870644. Warrendale, PA: Society for Automotive Engineers.

U.S. Department of Energy. 1996. "An Assessment of the Market Benefits of Alternative Motor Vehicle Fuel Use in the U.S. Transportation Sector: Technical Report 14, Market Potential and Impacts of Alternative Fuel Use in Light-Duty Vehicles: A 2000/2010 Analysis." DOE/PO-0042, Washington, DC: U.S. Department of Energy, Office of Policy and Office of Energy Efficiency and Renewable Energy. January 1996.

U.S. Department of Energy/Energy Information Administration. 2003. *Annual Energy Outlook*. Washington, DC: U.S. Department of Energy.

CHAPTER 15

The "Chicken or Egg" Problem Writ Large: Why a Hydrogen Fuel Cell Focus Is Premature

John M. DeCicco

A classic question in transportation energy policy takes the form of a chicken or egg problem, posed as, "What comes first, the vehicle or the fuel?" Partial answers to this dilemma were developed for earlier alternative fuels used in internal combustion engines. Addressing only the vehicle technology aspect of the broader sustainable transportation issue, the hydrogen fuel cell car is now held out as an ultimate solution to energy related concerns. Although the chicken or egg problem is quite daunting for hydrogen fuel cell car technology, its long term promise is earning it high-profile efforts to overcome the infrastructure and technical barriers inhibiting commercialization.

The paradigm shift to hydrogen rests on a set of inferences about the ultimate superiority of this technology. But a careful evaluation of these arguments suggests that such focus on hydrogen is premature at best. It also raises a deeper question, namely, whether a technological solution to transportation energy problems can really be found prior to, or even as a pathway toward, policies that measurably address the concerns that motivate such technology development. In short: What comes first, the solution to the problem or the commitment to solve it?

After examining the U.S. transportation energy policy debate, this chapter concludes that a strategic focus on hydrogen cars is not justified. Moreover, neither this technological solution nor any other is likely to be realized prior to a political decision to control the oil consumption and CO_2 emissions problems being used to justify the technology strategy.

Introduction

Transportation oil use poses perhaps one of the most daunting challenges for energy and environmental policy. Reasons for the difficulty include the uniqueness of the petroleum resource, the emotional value of the automobile, and the magnitude of the global institutions that have grown from and robustly sustain the symbiosis between oil and the automobile.

Incremental technology improvements seem to pale in comparison to the scale of transportation oil use and its rate of growth. Yet a generation's worth of efforts to develop workable alternatives has failed. As of 2000, alternative fuel use in the United States amounted to just 0.36 billion gallons, compared to 166 billion gallons of total highway petroleum fuel consumption (Davis and Diegel, 2002, Tables 2.3 and 2.4).

An incremental approach that has succeeded is the reduction in conventional air pollution from highway vehicles. Periodic tightening of standards has led to ongoing refinements of emissions controls, reducing the light duty vehicle air pollution inventory in spite of steady growth in driving (U.S. Environmental Protection Agency, 2000). Fully taking motor vehicles out of the equation for healthy air will involve a great deal of further effort: for diesel engines, for long-lived on-road stocks, and particularly for diffusing the best vehicle and fuel technologies to emerging markets apace with growth. The progress seen in this arena signals hope that the other problems might be solved.

Nevertheless, a fundamental physical issue limits how well lessons from air quality strategy can be applied to energy-related problems. While no laws of science limit the extent to which combustion can be cleaned up, energy consumption increases as more features treasured in the marketplace are incorporated into automobile design. Sophisticated controls and aftertreatment technology now enable automakers to build gasoline fueled internal combustion engines with near-zero emissions of conventional pollutants. Such decoupling of social harm from market value is not in sight for the oil dependence and global warming problems.

Or is it? Hydrogen, the "forever fuel" as it was dubbed two decades ago (Hoffmann, 1981), promises just such a decoupling. Hydrogen is promoted as the ultimately reliable, secure, clean, and carbon free energy carrier. The need for such a solution is especially acute for personal automobiles, which appear unsuited for extensive electrification. Such need begets hope, a hope that has been bolstered by the research progress in fuel cells, which are ideal energy conversion devices for hydrogen. International government–industry consensus now appears to pick hydrogen fuel cell technology as the long term winner, manifest by priority resource commitments for research, development, and demonstration (U.S. Department of Energy, 2002 and 2003b).

A Policy Paradigm Shift

This emerging strategic choice represents a tipping of the scales in a long running debate on transportation energy policy. Proponents of one technology and fuel option or another have argued that a strategic push toward a preferred alternative is essential for overcoming the high barriers to transition (e.g., Gray and Alson, 1985; Sperling, 1995). An opposing view is that only the market can make the right choices, and so government policies should be fuel neutral. Another similar view is that even though new fuels are clearly needed, the uncertainties dictate that public policy should promote a diversity of alternatives (Sperling, 1989, Chap. 17). A related notion is that fuel or resource diversification is a good cure for the alleged ills of a singular dependence on oil.

The twin principles of fuel neutrality and diversity have been enshrined in key federal policies of the past two decades, including the Alternative Motor Fuels Act (AMFA) of 1988 and the Energy Policy Act (EPACT) of 1992. While there are some notable exceptions to this basic respect of market decision making, the effects of non-neutral, strategically oriented policies also remain largely negligible, at least in the context of recent history.

One exception is fuel taxation policy. Motor fuel taxes range from a subsidy of \$0.82 per gallon of gasoline energy-equivalent for agriculturally derived ethanol to an added tax of \$0.22 per gallon of gasoline energy-equivalent for diesel fuel (\$0.18 per gallon of diesel itself). Moreover, at the time of this writing, a "renewable fuel" mandate to further promote ethanol received broad bipartisan support during Congressional deliberations on a new energy bill. A number of markets outside the United States have seen significant but still restricted use of some fuels, such as natural gas in several countries and ethanol for a time in Brazil, as a result of favorable tax treatment. These effects were more pronounced during periods of high oil prices. Nevertheless, because of the fundamental economic fact that petroleum is inexpensive in terms of basic resource cost, fuel pricing policy has served only to sustain subsidized niches and has had no impact on major energy sector investment decisions to date.

The other notable exception to fuel neutrality is California's ZEV mandate. Although it is *de jure* a performance standard, it was originally a *de facto* mandate for non-combustion technologies. Its target of promise had been the battery powered car. But the ZEV mandate is now substantially rationalized as paving the way for hydrogen fuel cell vehicles, which are foreseen as playing a prominent role according to the latest revisions of the rule (California Air Resources Board, 2003).

One reason for optimism about fuel cell cars, in contrast to battery electric cars, is that major automakers have embraced the technology. Indeed, automakers are cooperating with governments to promote, develop, and demonstrate fuel cell vehicles, in contrast to their grudging approach to battery electric vehicles. Nevertheless, it is important to note that automakers are supporting fuel cells as a solution that can come about without regulation. They favor a reliance on technology policy, partnerships, and market incentives, without mandates or performance standards. This view is not shared, of course, by most environmental and some government policymakers, for whom the regulatory driver of the ZEV mandate is still seen as essential for realizing the fuel cell vision.

The ZEV mandate—along with the California Fuel Cell Partnership, the new U.S. federal technology policies of FreedomCAR and Freedom Fuel, and similar initiatives announced in Europe and Japan—can be viewed as marking a paradigm shift away from market-based, fuel-neutral policies for managing transportation energy problems and toward a strategic push for a hydrogen fueled future.

Rationales for Change

A number of factors motivate a change in the automotive power train (DeCicco, 2001). The dominant factors are those related to public concerns. Other factors relate to consumer amenities, particularly new vehicle features that are likely to grow in importance.

Three rationales for change pertaining to societal concerns are generally within the purview of public policy. First, public health concerns imply a need to reduce emissions of criteria air pollutants in order to improve air quality. Second, global warming concerns imply a need to reduce emissions of greenhouse gases that cause climate disruption. Third, energy security worries lead to calls for reducing consumption of petroleum products that incur economic risks and security liabilities.

Two rationales for change pertain to market-driven, consumer-oriented features of motor vehicles. One is the value of power train electrification, enabling highly controllable, quiet, smooth, and customized responsiveness to driver input. The other is the need for ample onboard electric power to provide a variety of amenities as well as enable power train electrification.

These last two factors are extensions of traditional customer satisfaction issues of comfort, performance, quality, reliability, and affordability. Automakers evaluate all of these factors within the range of design issues reflecting other societal concerns, such as safety and accessibility, as well as market desires, including cultural meanings and expectations of the vehicle.

These considerations will influence the choices made as the industry plans its products in a shifting market and policy context. Not all factors are equally important at a given point in time, and so their relative emphasis also determines the choices made. Any option for future vehicle fuel design, including a default slow evolution of existing technology, must compete in order to command the substantial investments entailed in both product development and infrastructure provision.

It is in light of these rationales for change that the argument for a strategic focus on hydrogen fuel cell vehicles has developed. The key elements in this argument are that:

- Many options exist for alternative vehicle-fuel technologies that can address one or some of the factors motivating change.
- A superiority in addressing a single factor is not sufficient to push any given technology to the fore.
- Acceptable technologies must ultimately meet market dictated cost constraints; superiority in addressing societal factors is unlikely to justify perpetual subsidy.

It is because hydrogen fuel cells might one day meet all of the requirements for the automotive power train of the future that they have become such a focus of government and industry initiatives.

Evaluating the Strategic Choice

While the above reasoning provides a justification for the hydrogen fuel cell strategy, it is important to take a dispassionate look at the arguments within the context of policy issues, market drivers, and technological competition.

Public Health

The need to reduce the health impacts of air pollution has provided much of the impetus for fuel cell vehicles. By the late 1980s, many people were discouraged about the ability to achieve adequate and durable pollution control of gasoline vehicles. In-use emissions greatly exceeded the standard levels, with control effectiveness degrading as vehicles aged. Even relatively new vehicles sometimes had excessive emissions in real-world usage. The need for radically different technology was acutely felt in the Los Angeles area, which had the nation's worst air pollution and vehicle use that was increasing at a rate nearly twice that of population growth.

Some policymakers came to view it as essential to move beyond petroleum to attain healthy air. Stringent emissions control requirements were premised on the use of fuels, such as methanol or natural gas, held to be inherently cleaner than petroleum fuels. Strict standards were issued in spite of uncertainty about whether they could be met by improved gasoline vehicles. Moreover, concerns about emissions control durability and the persistence of high emitters suggested that onboard combustion would be ultimately inadequate, creating an impetus for zero-emission vehicle (ZEV) technology (Lloyd and Leonard, 1992). Analysts projected that battery electric vehicles (BEVs) would cost effectively reduce criteria emissions compared to the best gasoline technology then foreseen (DeLuchi *et al.*, 1989). While most energy specialists initially called for BEV research, subsidization, and promotion, General Motors' January 1990 display of the Impact concept car inspired the ZEV mandate.

California researchers concluded that electric vehicle commercialization no longer depended on technical breakthroughs. While acknowledging battery limitations, they projected acceptance by many consumers and significant EV market shares. Fuel cells would be a further technological step that would then enable widespread use of combustion free vehicles. The ZEV mandate was also seen as necessary for ensuring air quality attainment in a 2010 time horizon. Without electric vehicles, for example, nitrogen oxide (NO_x) emissions ten times higher than now-planned low emission (LEV) II levels, similarly high levels of hydrocarbons and other pollutants, and a fuel sulfur level of 300 parts per million (ppm), were projected for the California light vehicle stock in 2010 (Wang *et al.*, 1990).

In the 1990s, however, gasoline emissions control technologies advanced well ahead of expectations. Current LEVs are already cleaner than the projections for 2010 made in the late 1980s. Both California and the federal government are phasing in the more stringent low emission vehicle (LEV) II and Tier 2 programs beginning in 2004. Fuel sulfur levels are being

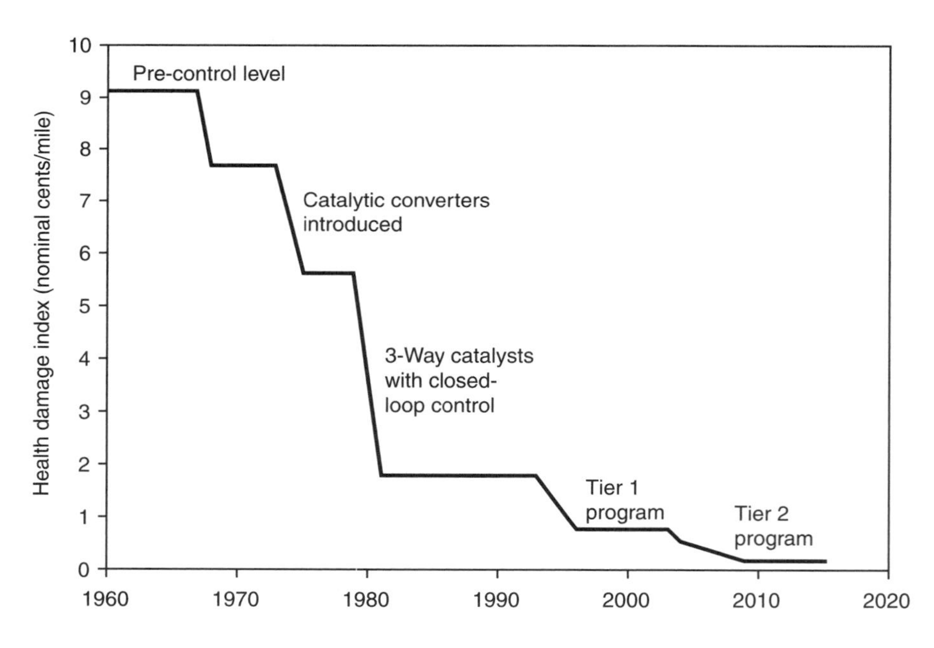

FIGURE 15-1. Progress in regulating air pollution from new U.S. passenger cars. The health damage index was derived from U.S. federal passenger standards for car emissions by weighing NO_x, HC, and CO emissions using median damage cost factors from Delucchi (1997). These nominal factors, based on regulatory standard values, do not account for changes in population exposure, vehicle mix, and emissions control reliability over the 60-year period illustrated.

cut to an average of 30 ppm nationally and below 10 ppm in California by 2010. Figure 15-1 illustrates the progress in conventional vehicle emissions reduction since the precontrol era of the 1960s through the Tier 2 program over the coming decade.

Attainment of air quality goals over the next two decades is no longer dependent on ZEV technology. Thus, the public health need for technologies such as fuel cells may be pushed forward to a long term, roughly year 2030, timetable. More stringent standards may be needed to address issues, such as ultrafine particulate matter and toxics, poorly addressed by currently planned emissions control programs. Long term control durability and the problems of high emitters, particularly among older vehicles, also remain as concerns. On the other hand, the LEV II and Tier 2 standards do not represent the limit of gasoline emissions control. For example, one automaker has already demonstrated "zero level emissions vehicle" (ZLEV) technologies that take gasoline vehicle emissions to even lower levels (Honda, 2000).

Global Warming

Although a U.S. policy to require cuts in greenhouse gas (GHG) emissions has not been established, the need to address global warming is now widely acknowledged. International support is strong for the Kyoto protocol. Several states have adopted policies or plans to reduce GHG emissions, including the 2002 California law requiring cost effective reductions in GHG emissions from light duty vehicles.

The most well defined policy to date for reducing automotive GHG emissions is the 1998 voluntary agreement to cut average new European passenger car CO_2 emissions 25 percent by 2008 (Association des Constructeurs Européens d'Automobiles, 2002). This level of reduction assumes refinement of petroleum fueled vehicles, including increased diesel share. Although the program's progress stalled in 2002, a 12 percent reduction was achieved over the first 4 years of the agreement (de Saint-Seine, 2003).

The fact that significant global reductions are expected raises the importance of global warming in vehicle strategy. Nevertheless, absent policy change, new U.S. light duty fleet fuel consumption and CO_2 emission rates may remain on the upward trend they have shown over the past 15 years, as illustrated in Fig. 15-2.

The long term promise of hydrogen for addressing global warming rests on its uniqueness as a carbon free chemical fuel. The only other carbon free energy carrier now considered potentially feasible is electricity. Other alternatives for a very low or zero-net carbon vehicle fuel system entail renewable energy, i.e., using agricultural cycles that recycle carbon or fossil fuels with carbon sequestration. Similar resources can also be used to produce hydrogen. As is the case for electricity, the impact of hydrogen on global warming depends on how it is produced. Two questions are involved. One is whether hydrogen is unique enough to make it the most likely long term

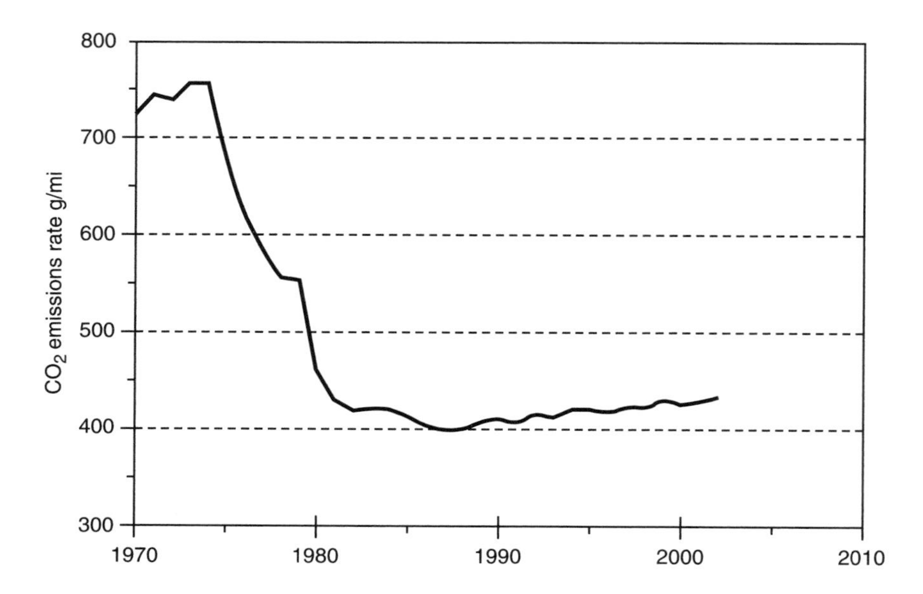

FIGURE 15-2. New fleet average CO_2 emission rates of U.S. cars and light trucks. *Source:* Nominal CO_2 emissions rate derived assuming fixed 15 percent fuel economy shortfall and 8800 g/gal CO_2 emissions factor using statistics from Hellman and Heavenrich (2003).

option. The other is how a strategic push toward hydrogen rates against other options for near and medium term GHG reduction.

Given current knowledge, a special case for hydrogen is difficult to make in light of the potential for renewable fuel cycles that can produce convenient liquids (Lynd *et al.*, 1995). Hydrogen's long term appeal is stronger when considering fossil or nuclear primary resources because its carbon free nature then pits it only against electricity. In any case, given how long it will take to develop an ultra-low GHG emissions system of any sort, this aspect of the case for hydrogen remains hypothetical.

For near term vehicular GHG reduction, combustion-based power trains (conventional as well as hybrid electric) that use improved gasoline or diesel are the only serious options. Although any new fuel involves infrastructure lags, further reductions in GHG intensity can be had in the medium term using various reduced or low-carbon fuel cycles, still using combustion-based power trains. Moreover, as GHG control policies evolve, the technology assessments used to justify the reduction targets will tend to be conservative, so that the risks and costs imposed on the industry are manageable. This inherent policy conservatism is the norm; deviations from it are unlikely to be robust, as California's ZEV mandate experience demonstrates.

Therefore, demonstrable cost effectiveness looms large as an evaluation criterion for near- and medium-term strategies (e.g., National Research Council, 2002). Progress along such lines is probably the most that can be hoped for over the next two decades, because the United States has yet to really pursue even "no regrets" strategies, which pass only private cost effectiveness tests, let alone socially cost effective measures. This situation imposes very high hurdles for any technology that requires fundamental breakthroughs for mass-market workability, even when a viable transition path is clear. Such is certainly not the case for the hydrogen car. Thus, many other options will out-compete it as the policymakers enter a new realm of GHG mitigation.

Energy Security

Controlling petroleum consumption and developing alternatives to oil dependence have been stated goals of U.S. energy policy for nearly 30 years. This public concern is borne of the experience of economic, military, and other security risks associated with control of access to petroleum resources.

Since the time of the oil shocks of the 1970s, major shifts have occurred in the structure of oil use in the U.S. economy. Sectors that could switch to other fuels, such as natural gas, have largely done so. Nontransportation oil use is now lower than it was in 1970, in spite of substantial economic growth (Davis and Diegel, 2002, Table 1.12). Much of the remaining use is for lower value products whose usage is growing slowly at most and which do not drive the energy market in the way that high-value, steady-growth transportation fuels do. The transportation sector remains almost wholly dependent on petroleum as a source of energy. The sector now accounts for two-thirds of U.S. petroleum use, compared to one-half in 1970 (Davis and Diegel, loc. cit.). Even in the mature U.S. market, fuel consumption has grown steadily since the effect of the corporate average fuel economy (CAFE) standards played out by the early 1990s.

During Congressional testimony in 1980, Daniel Yergin stated that "one out of nine barrels of oil used in the world every day is burned as gasoline on the American highway" (Yergin, 1980). That statement remains true today (Davis and Diegel, 2002, Tables 1.4 and 2.4). It is a striking testament to how growth in this largest single market for high-value petroleum products—even net of the significant step-up in auto efficiency that followed the oil shocks—has kept pace with both structural changes in the composition of energy use and the more rapid relative growth in less mature auto markets throughout the world. Meanwhile, "automobilization," both as a means of transportation and as a pillar of economic development, plays a key role in the aspirations of other nations of the world. This ongoing trend ensures steady growth in global oil demand. The resulting pressure on the oil market will only aggravate the factors that lead to energy security concerns about oil dependence.

While concern about oil dependence enjoys widespread, bipartisan rhetorical support, a certain wariness is in order given the history of U.S. transportation energy policy. On the heels of an initial strong push for conservation came a short-lived but expensive dabbling in synthetic fuels, followed by the alternative fuels emphasis that has dominated this area of federal policy since the late 1980s. Over the years, alternative fuel proponents have articulated assumption-driven scenarios regarding the promise and plausible long term success of their chosen fuel and technology options.

Under the banner of energy security, such scenarios have rationalized research and development programs, subsidies, demonstrations, and even limited regulatory policies such as the CAFE credits of AMFA and alternative fuel fleet provisions of EPACT. However, given the magnitude of the transition barriers, enabling any new vehicle-fuel technology to truly compete with established technology would require much stronger government policy and incur substantial costs (see Chapter 14). To date, no alternatives have been compelling enough to justify the level of effort needed for market transformation.

Proponents argue that the multiple benefits of hydrogen fuel cells are what it will take to finally make a compelling case to move beyond oil. They point to the ability to produce hydrogen from diverse feedstocks as a key aspect of its appeal. These arguments were articulated as early as 1987 by Paul Werbos (1987) and more recently by a number of technology proponents (e.g., MacKenzie, 1994; Mark, 1996). Such reasoning partly underpins the FreedomCAR and Freedom Fuel initiatives. However, hydrogen is not unique in this regard; many benefits of alternative fuels are independent of the type of power train. The point here is not to argue for or against a particular fuel choice. Rather, it is to note that it is far from clear how hydrogen fuel cells, which face enormous technical and cost barriers, might break out of the box that has trapped other promising alternatives, none of which have made headway in spite of less daunting barriers (see Chapter 9).

The hydrogen focus thus seems little different than past failed attempts to advance a technical solution to the problems of U.S. transportation oil dependence. The fact that a more extensive government/industry research initiative has coalesced around hydrogen does not negate the fundamental limitations of any "promising solution" type of strategy. What it does do is highlight in stark relief another apparent government/industry consensus, as least as far as U.S. federal policy is concerned, namely, that of not pursuing, to any significant degree, more accessible options for measurably managing transportation oil demand. As have all recent editions of the report, the DOE's 2003 *Annual Energy Outlook* projects U.S. petroleum consumption steadily rising and increasingly dominated by transportation (U.S. Department of Energy, 2003a). Only if there were no other options for bending this growth curve would considerable effort on a long-term, high-risk strategy warrant such focus.

Market Factors

Market forces are not a significant driver for environmental improvements in motor vehicles, and consumer interest in fuel economy remains very weak. On the other hand, efficiency and environmental friendliness may be emerging as future market factors. Several studies have shown growth potential for hybrid vehicles (J.D. Powers, 2002). Major automakers now promote the environmental attributes for some of their products as part of corporate image campaigns. Nevertheless, the customer value placed on high efficiency or low emissions seems likely to remain too weak to motivate major product design changes for the foreseeable future.

A more influential market driver is the growth of electric and electronic systems on vehicles. The demands come not only from a variety of comfort, convenience, communications, and entertainment features, but also from an evolution to electromechanical operation for many auxiliary devices that to date have been purely mechanical, such as power steering, valve train actuation, braking, suspensions, and air conditioning, among others. Moreover, power train electrification enables new levels of driver responsiveness and controllability in addition to fuel economy benefits.

Efficiently meeting higher electric loads requires voltages higher than 12 volts, with the initial step being 42-volt technology. This trend will raise the value of high capacity onboard power generation, of which the fuel cell is perhaps the ultimate solution. However, competing technologies—ranging from stepped-up conventional alternators through integrated starter–generators to hybrid power trains—will serve growing onboard power needs for many years to come.

Thus, the weak but emerging importance of fuel economy and environmental friendliness, along with the growing importance of electric capacity, can serve as market drivers for automotive fuel cells. However, even if the technology's technical and infrastructure barriers are overcome, these drivers are unlikely to make the business case compelling over the next two decades because lower cost, less disruptive technologies can meet such needs for the foreseeable future (DeCicco, 2001).

Conclusion

A priority focus on hydrogen fuel cells for automobiles is difficult to justify based on what is known of this technology when it is evaluated against competing options and the forces likely to shape automotive design over the coming decades. Thus, initiatives like FreedomCAR and the California Fuel Cell Partnership are premature, even though some level of steady, long term research is justified.

A notable aspect of the situation is the uneven status of the societal concerns that motivate hydrogen cars. Hydrogen might be an ideal joint solution to the problems of air quality, global warming, and energy security.

But the roles that these concerns have in public policy are not now aligned to require a joint solution, and they may not become so aligned any time soon.

Air quality is an established policy driver. Indeed, it has been so effective in inducing technology change that it is now serving to secure the competitiveness of gasoline internal combustion engine technology against its challengers on public health grounds. Concerns have been raised that U.S. air quality regulations may unduly constrain diesel technology and its GHG reduction benefits. Nevertheless, advances in diesel emissions control may well be able to keep pace with the rate at which "dieselization" of the U.S. market is needed to meet global warming concerns that are only slowly translating to reduction targets.

Global warming wasn't even on the public policy radar screen a generation ago when both air quality and energy security became major issues. Although it is now a major concern, GHG reduction requirements are likely to evolve gradually, perhaps following the air quality model, and so are unlikely to require radical technology change anytime soon. The ZEV mandate experience, even though it is now linked to the hydrogen fuel cell vision, serves as a cautionary tale regarding the limits of technology-forcing regulatory policy. Requirements for zero-GHG vehicles seem quite far off, if not implausible, over the generation to come.

Energy security remains a perennial concern as the United States marks the 30th anniversary of the 1973 oil embargo. However, most U.S. policymakers seem satisfied with established oil supply strategies, ranging from enhanced and unconventional production technology to strategic options for managing the geopolitical risks. Thus, political support has been insufficient to enact new policies that would significantly affect U.S. petroleum consumption. As for strategies that have promoted the many promising nonpetroleum automotive fuels, the evidence indicates that their effectiveness to date can be summed up as "The Alternative Fuel Follies," to paraphrase the title of a prescient *Popular Science* article (McCosh and Brown, 1992).

Therefore, the most important thing that has to happen to increase the prospects for hydrogen fuel cell cars (or any breakthrough vehicle-fuel solution) is to better define the requirements for addressing the concerns being used to justify the technology development. Either global warming or energy security (or both) will need to be taken seriously enough that targets for measurable progress are established, analogous to those that have reshaped technology for reasons of air quality. This step is a political one that is not likely to be hastened as long as the public discussion is focused on technical steps that are supposedly necessary first.

In the tenacious realm of transportation energy use, success has not been seen over many years of attempting to develop technical solutions as a way to facilitate policies that might then guide beneficial change. Political commitment to cut oil use or GHG emissions seems essential before any new technological solution, no matter how promising on paper, can come to fruition. Moreover, if such a commitment is made, the resulting search for ever better solutions is likely to yield options far superior to those hypothesized today in a context where the need for real progress remains remote.

References

Association des Constructeurs Européens d'Automobiles. 2002. *ACEA's CO_2 Commitment: A 35 Million Tonnes CO_2 Kyoto Contribution to Date*. Bruxelles: Association des Constructeurs Européens d'Automobiles.

California Air Resources Board. 2003. *ARB Modifies Zero Emission Vehicle Regulation*. News Release. Sacramento, CA: California Air Resources Board, April 24, 2003.

Davis, S. C., and S. W. Diegel. 2002. *Transportation Energy Data Book, Edition 22*. Oak Ridge, TN: Oak Ridge National Laboratory, September.

DeCicco, J. M. 2001. *Fuel Cell Vehicles: Technology, Market, and Policy Issues*. Warrendale, PA: Society of Automotive Engineers.

Delucchi, M. A. 1997. *The Annualized Social Cost of Motor-Vehicle Use in the United States, 1990–1991: Summary of Theory, Data, Methods, and Results*. Davis, CA: University of California, Institute of Transportation Studies, June.

DeLuchi, M., Q. Wang, and D. Sperling. 1989. Electric Vehicles: Performance, Life-Cycle Costs, Emissions, and Recharging Requirements *Transportation Research* 23A(3): 255–278.

de Saint-Seine, S. 2003. EU Automakers Won't Meet 2008 CO_2 Goals. *Automotive News Europe* (Nov. 17): 1.

Gray, C., and J. Alson. 1985. *Moving America to Methanol*. Ann Arbor, MI: University of Michigan Press.

Hellman, K. H., and Heavenrich, R. M. 2003. "Light-Duty Automotive Technology and Fuel Economy Trends, 1975 through 2003." Report EPA 420R-03-006. Ann Arbor, MI: U.S. Environmental Protection Agency, Office of Transportation and Air Quality, Advanced Technology Division, April.

Hoffmann, P. 1981. *The Forever Fuel: The Story of Hydrogen*. Boulder, CO: Westview Press.

Honda. 2000. "Honda Develops Gasoline Engine With Zero-Level Emissions." Press Statement. Tochigi, Japan: Honda Motor Co., Ltd., October 20.

J. D. Powers. 2002. "Consumers Want Hybrid Models in Virtually All Vehicle Segments." News Release on Hybrid Vehicle Consumer Acceptance Study. Los Angeles, CA: J.D. Powers & Associates, March 6.

Lloyd, A. C., and J. Leonard. 1992. "The Role of Battery and Fuel Cell Powered Electric Vehicles in Air Quality Planning." Presented at the Transportation Research Board, 71st Annual Meeting, January.

Lynd, L. R., R. T. Elander, and C. E. Wyman. 1995. "Likely Features and Costs of Mature Biomass Ethanol Technology." Paper Presented at the 17th Symposium on Biotechnology for Fuels and Chemicals, Vail, CO, May.

MacKenzie, J. J. 1994. *The Keys to the Car: Electric and Hydrogen Vehicles for the 21st Century*. Washington, DC: World Resources Institute.

Mark, J. 1996. *Zeroing Out Pollution: The Promise of Fuel Cell Vehicles*. Cambridge, MA: Union of Concerned Scientists.

McCosh, D., and S. F. Brown. 1992. The Alternate Fuel. Follies *Popular Science* (July): 54–59.

National Research Council. 2002. *Effectiveness and Impact of Corporate Average Fuel Economy (CAFE) Standards*. Washington, DC: National Academy Press.

Oak Ridge National Research Laboratory, 2002. *Transportation Energy Data Book, Edition 22*. Oak Ridge, TN: U.S. Department of Energy, September.

Sperling, D. 1989. *Alternative Transportation Fuels: An Environmental and Energy Solution*. New York: Quorum Books.

Sperling, D. 1995. *Future Drive: Electric Vehicles and Sustainable Transportation.* Washington, DC: Island Press.

U.S. Department of Energy. 2002. *A National Vision of America's Transition to a Hydrogen Economy—to 2030 and Beyond.* Washington, DC: U.S. Department of Energy, February.

U.S. Department of Energy. 2003a. "Annual Energy Outlook 2003, with Projections to 2025." Report DOE/EIA-0383. Washington, DC: U.S. Department of Energy, Energy Information Administration, January.

U.S. Department of Energy. 2003b. FreedomCAR and Vehicle Technologies Program, Washington, DC: U.S. Department of Energy, March.

U.S. Environmental Protection Agency. 2000. "Control of Air Pollution from New Motor Vehicles: Tier 2 Motor Vehicle Emissions Standards and Gasoline Sulfur Control Requirements." Issued by the U.S. Environmental Protection Agency. Washington, DC: Federal Register 65(28): 6698ff., Feb. 10.

Wang, Q., M. A. DeLuchi, and D. Sperling. 1990. Emission Impacts of Electric Vehicles. *Journal of the Air Pollution and Waste Management Association* 40(9): 1275–1284.

Werbos, P. 1987. "Oil Dependency and the Potential for Fuel Cell Vehicles." SAE Paper No. 871091. Warrendale, PA: Society of Automotive Engineers.

Yergin, D. 1980. Testimony to the Committee on Energy and Natural Resources, Hearing on the Potential for Improved Automobile Fuel Economy between 1985 and 1995. Washington, DC: U.S. Senate, April 30.

CHAPTER 16

The Case for Battery Electric Vehicles

Paul B. MacCready

The case for battery electric vehicles (BEVs) has been greatly enhanced by significant recent improvements in lithium batteries. New lithium batteries have about three times greater energy density than nickelmetal hydride (NiMH) batteries used in the recently, but no longer, available EV-1 from General Motors and RAV4-EV from Toyota. Recent progress with lithium technology makes battery electric power an increasingly viable source of car propulsion energy. Some lithium batteries now store nearly 200 Watt hours per kilogram (Wh/kg) of energy and have very high efficiency charging and discharging cycles. With a bidirectional charge-discharge hookup, the batteries can also be used to provide off-board power to help an electric utility adjust its grid capacity and handle peak loads on hot days.

My confidence in lithium battery technology is based on my long experience with it at AeroVironment, Inc. I also am well acquainted with hydrogen fuel cell vehicle systems. To our knowledge, our tiny Hornet plane is the first fuel cell powered airplane. We have integrated and used fuel cells for large and small systems, and are thoroughly conversant with their potentials and challenges. It is this familiarity with battery powered cars and airplanes, with fuel cells, and with combined systems that moves us to explore the reasonable use of the remarkable new lithium batteries for cars. We will start with their use in the hybrid electric vehicles until the economics, convenience, and energy potential permit the lithium batteries to do the whole job.

A principal goal at AeroVironment, Inc., has been to provide technology for air, land, and water transportation using minimal amounts of energy and preferably entirely nonpolluting technologies. We have developed and used devices employing human power; turbulence and wind and wind shear; solar power; gasoline, diesel, and natural gas internal combustion engines (ICEs); capacitors and supercapacitors; and fuel cells and battery electric power technologies.

Some of the advanced vehicles developed over the years at AeroVironment include:

Gossamer Condor and Gossamer Albatross human power airplanes
Solar Challenger, a solar powered airplane
Sunraycer, a solar and battery powered land vehicle
GM Impact battery powered car
Helios, the solar power airplane that has climbed to 96,863 feet, or 2 miles higher than any plane had flown continuously from the ground
Pointer UAV, a 9-pound drone airplane
Hornet, a 6-oz drone with a 14-inch wingspan that flew in March 2003 solely on fuel cell power

The Pointer, a 4-pound Raven, and a 6-pound Dragon Eye have already become commercial products.

Thus, we have worked on many propulsion devices, always with an eye toward long term environmental suitability. For the problem of car transportation, one possible solution now appears to us to be battery power—because the new lithium batteries are so good and will certainly be even better in the future. This is just one solution for personal mobility on land. As urbanization inevitably increases, attention will have to be paid to all possible solutions—for buses, trains, scooters, and ride sharing, and using computer interfaces to minimize personal travel.

A vehicle that gets its electricity solely from a bank of lithium batteries is an excellent energy strategy to address the multidimensional problems of oil dependence and pollution inherent to the world's current transportation system.

This chapter addresses the question of how the United States might create a transportation system with dramatically lower oil use and greenhouse gas emissions. It concludes that the lithium battery powered car now represents a better energy strategy to achieve an economically viable solution than hydrogen fuel cells do.

The Challenge of Sustainability

Most U.S. adults have used cars powered by fossil fuels through their entire lifetimes. The contemplation of any other surface transportation method has received relatively little attention. Yet, oil use and greenhouse gas emissions continue to increase. We are moving directly and quickly toward a significant global challenge. It is now logical to explore other possible solutions to reduce fossil fuel use while accommodating increased demand for more transportation.

Figure 16-1 shows most energy uses on earth except for nuclear, tidal, and geothermal power. All the energy shown is renewable. The top six categories are renewable within a human lifespan, but the bottom category, including

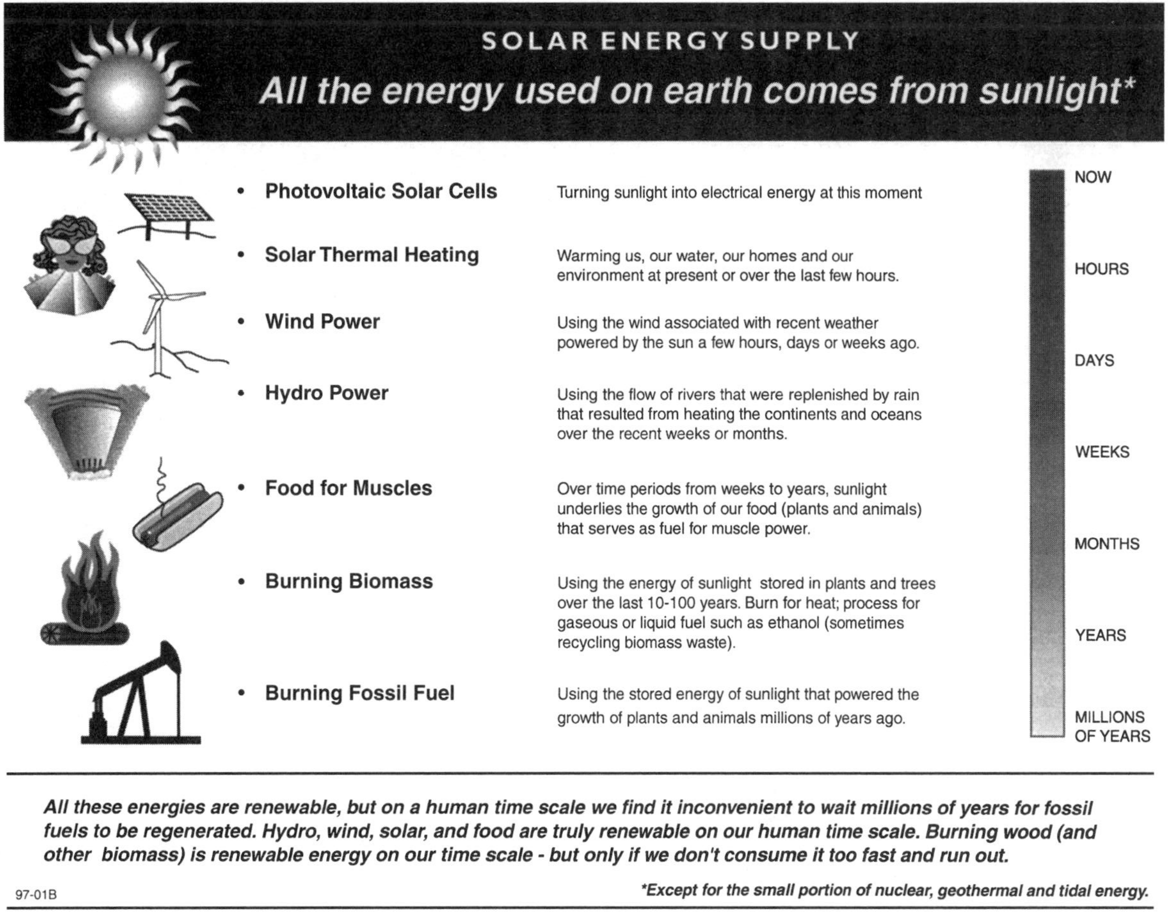

SOLAR ENERGY SUPPLY

*All the energy used on earth comes from sunlight**

Source	Description	Time scale
Photovoltaic Solar Cells	Turning sunlight into electrical energy at this moment	NOW
Solar Thermal Heating	Warming us, our water, our homes and our environment at present or over the last few hours.	HOURS
Wind Power	Using the wind associated with recent weather powered by the sun a few hours, days or weeks ago.	DAYS
Hydro Power	Using the flow of rivers that were replenished by rain that resulted from heating the continents and oceans over the recent weeks or months.	WEEKS
Food for Muscles	Over time periods from weeks to years, sunlight underlies the growth of our food (plants and animals) that serves as fuel for muscle power.	MONTHS
Burning Biomass	Using the energy of sunlight stored in plants and trees over the last 10-100 years. Burn for heat; process for gaseous or liquid fuel such as ethanol (sometimes recycling biomass waste).	YEARS
Burning Fossil Fuel	Using the stored energy of sunlight that powered the growth of plants and animals millions of years ago.	MILLIONS OF YEARS

All these energies are renewable, but on a human time scale we find it inconvenient to wait millions of years for fossil fuels to be regenerated. Hydro, wind, solar, and food are truly renewable on our human time scale. Burning wood (and other biomass) is renewable energy on our time scale - but only if we don't consume it too fast and run out.

97-01B

Except for the small portion of nuclear, geothermal and tidal energy.

FIGURE 16-1. Solar energy supply.

the fossil fuels, requires millions of years to renew and so for humans must be considered nonrenewable.

Fossil fuel, including the oil that powers transportation, has been ubiquitous to the point that we have become accustomed to thinking of it as permanently available. It is not. U.S. oil production peaked in 1970 and has dropped by roughly one-quarter since then. Some predictions put peak global oil production within this decade, while other predictions see the peak coming a decade later. In any case, consumption releases CO_2, which virtually all atmospheric scientists see as materially affecting Earth's weather. Coal and natural gas will not run out so quickly, but they too contribute to the release of CO_2. Solutions to the CO_2 challenge must either show there really is no problem from the increased atmospheric CO_2, somehow capture and sequester released CO_2, or prevent its discharge into the atmosphere in the first place.

Figure 16-2, showing the global mass of vertebrate life on land and in the air, provides a warning picture of human actions. In 2000, when this figure was prepared, the line indicating humans + livestock + pets was at 98 percent, leaving just 2 percent for natural wildlife. Ten thousand years ago, humans + livestock + pets were under 0.1 percent! The biggest changes

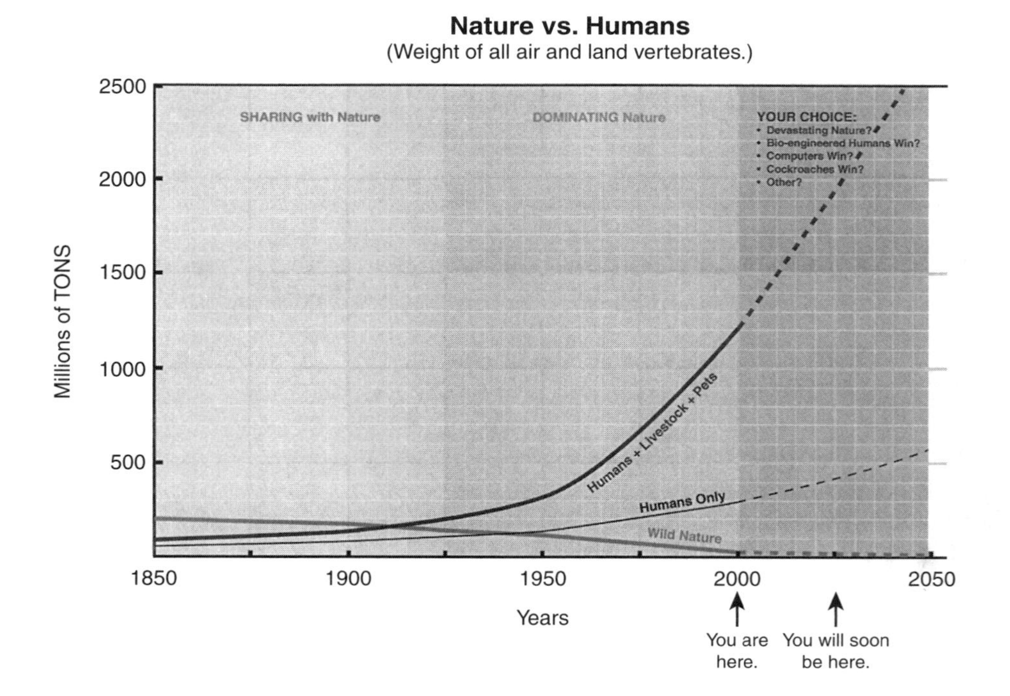

FIGURE 16-2. Global mass of vertebrate life on land and in the air.

started occurring 200 to 300 years ago as we initiated major digging for coal, then grew more dramatically as we explored for oil starting about 100 years ago. The details are actually much more complex, but the meaning is clear: Humans won as their abilities let them extract energy from the earth that was not renewable within the lifetime of civilization. This figure shows how humans dominate, and it also shows that we must make rational decisions now based on our "ownership" of the whole planet.

The message in Figs. 16-1 and 16-2 is that we must do more work with less fossil fuel energy—much less. We can obtain much more energy from natural, renewable resources, and we can use it much more effectively. The question is "Since we must, will we?"

Lithium Batteries as a Solution

BEVs with lithium batteries are similar to hydrogen fuel cell electric cars. In both, the primary energy source may be the same. Energy sources include oil, coal, natural gas, nuclear, hydropower, wind, and direct solar energy. These sources can power both types of vehicles when converted to an energy carrier, which could be electricity or hydrogen. Fuel cell vehicles use hydrogen as well as electricity energy carriers, but in both cases the end point is the same: mechanical energy delivered to the drive wheels. Both vehicles use electric motors and power electronics after the primary energy source passes through an electrochemical step. It is also true for both vehicle types that the possibility of avoiding oil use exists.

Because of the complexity of fuel cell systems, the large number of components required, and the issues of hydrogen storage, I find myself drawn to lithium batteries when we consider the comparison of start-to-end energy efficiency, range, performance, life cost, and convenience for cars. On the other hand, hydrogen may emerge for defined-use applications. For the 65,000-ft flight of our large, slow aircraft at high northern latitude in midwinter, present-day photovoltaic cells will not store enough energy to let the plane make it through the night at altitude. Therefore, liquid hydrogen appears to be the suitable fuel, permitting flights of 10 days or more solely on the hydrogen and atmospheric oxygen.

Stimulated by the market, lithium batteries have become widely used for cell phones and microcomputers, and they are taking on an important role in the tiny market of model airplanes. The available small lithium cells mostly offer 160 to 180 Wh/kg; the best are 220 Wh/kg. This is three times the energy of NiMH batteries and six times more energy per kilogram than lead acid cells. Power delivered is improving—now roughly 600 W/kg—and some new cells are even indicating much higher values.

BEVs with lithium batteries are presently showing well-to-wheel energy efficiencies between 1.5 and 4.0 times better than gasoline powered vehicles or fuel cell vehicles, depending on the primary energy source, and even much more when considering regeneration during braking. Vehicle driving ranges in excess of 300 miles are now possible using off-the-shelf lithium batteries. Automotive research indicates that vehicle acceleration from 0 to 60 mph is possible in 6 seconds or less with lithium battery packs that weigh up to 20 percent of the total vehicle weight. If a vehicle is charged in a home garage, personal inconvenience time may be less than with a conventional gasoline car or with a hydrogen fuel cell vehicle. Brooks (2002a), now with AeroVironment, reviewed the potential for operation with lithium batteries with very positive conclusions.

Life costs for lithium battery vehicles may soon be comparable to gasoline vehicles. Costs can be expected to decrease markedly when lithium batteries specifically tailored for cars are in wide use. In addition, lithium battery power offers substantial regenerative braking energy recovery. Recovery of this energy can extend the vehicle driving range in typical urban driving by 50 percent.

When used with a bidirectional charging setup, lithium batteries offer stability of electric utility current supply because they can continually and rapidly charge and discharge small amounts to a grid regulation ancillary service (Brooks, 2002b). Bidirectional charging can also use the surplus battery charge not needed for daily driving to supplement the utility grid needs during high loads on hot afternoons and evenings. After this loan of electricity back to the utility, the total drain would be partially or fully made up by full charging at night. Bidirectional charging thus offers vehicle owners the potential to generate value from the vehicle while parked, which typically occurs for 23 hours per day. This might cover all energy costs for driving, up to paying for a significant fraction of total vehicle cost.

As another example of the use of lithium batteries, consider the Antares sailplane made by Lange Flugzeugbau GmbH (Birch, 2003). The plane carries 76 kg of Saft's high energy rechargeable lithium batteries, which represent 13 percent of the vehicle's 600-kg all-up weight (including pilot, parachute, and water ballast). The batteries power a 2-meter diameter propeller connected to a 42-kW brushless electric motor. The motorpropeller system folds into the fuselage during glide. Assuming a 60-mph glide speed, this high performance plane could achieve 112 miles by the battery. The vehicle climbs to 3000 meters in 12 minutes, then glides for 100 minutes. The total battery consists of 72 cells in series, producing a nominal voltage of 260 volts, and has a capability of delivering 134 Wh/kg of battery weight. This battery is designed for 1500 charge–discharge cycles and is expected to have a service life of around 11 years.

Finally, there is the new lithium battery "tzero" car of AC Propulsion. This sports car was driven by the AC Propulsion chairman, Alan Cocconi, for 250 miles, much of it at 70 mph in tests in August 2003. It finished the trip with 15 percent of the range remaining. The 890-kg vehicle employs 6800 of the 18650 type lithium cells that offer 170 Wh/kg. Each cell weighs 43 g, for a total weight of 292 kg. The discharge–charge ratio turned out to be very high, supporting the regeneration mode. Long life is expected for the vehicle's power system, but this has not yet been verified. The tzero has been clocked at 3.6 seconds for acceleration from 0 to 60 mph.

It is helpful to start with the ultimate and then back off to the achievable when analyzing the potential for lithium powered vehicles. Here is the ultimate. If an electric car had 100 percent efficiency of electricity utilization, zero tire and aerodynamic drag, and 100 percent energy recovery from braking, its propulsion efficiency would be perfect. Except for the electricity used for control, instruments, radio, and lights, the range would be infinite.

Now look at the achievable. In/out lithium recharging is high, typically 93 percent for 1-hour charge/discharge rates. Tire drag in BEVs has been very low. The tires in the EV-1, for example, had drag of 0.7 percent of the vehicle weight, and similar special tires were used in the RAV4-EV. The aerodynamic drag of the EV-1 is also low, and eventually one should be able to maintain laminar flow over more of the body, as shown by research on suction on sailplane wings. Regenerative braking can be increased to over 80 percent net efficiency. When one starts with a very efficient battery system, such as lithium, and makes all the operating systems as efficient as possible, the final efficiency can be remarkable.

Because of the present high costs, a reasonable initial deployment strategy should be with a hybrid electric vehicle (HEV) having a battery electric power range of 75 to 100 miles. Use of battery electricity obtained from the grid through recharging should be able to handle about 90 percent of the annual driving distance of such a car, while using gasoline or an alternative fuel to augment power for longer trips. As battery costs decline and performance increases, the battery range can be extended and the internal combustion engine (ICE) contributions decreased or eliminated.

Batteries in cars offer a wonderful future. Commercial production and sale of battery powered cars, or hybrid cars with the battery assuming the main transportation job, offer the best known potential for environmentally sound transportation devices in the future.

References

Birch, Stuart. 2003. Batteries from Airlines to Autos. *Aerospace Engineering* (Jan/Feb): 13.

Brooks, Alec N. 2002a. "Perspectives on Fuel Cell and Battery Electric Vehicles." Report presented to the California Air Resources Board and the California Energy Commission. Sacramento, CA, December 10.

Brooks, Alec N. 2002b. "Vehicle-to-Grid Demonstration Project: Grid Regulation Ancillary Service with a Battery Electric Vehicle." Report presented to the California Air Resources Board and the California Energy Commission. Sacramento, CA, December 10.

CHAPTER 17

Hydrogen Hope or Hype

Daniel Sperling and James S. Cannon

Repeated efforts have been made to introduce alternatives to petroleum and internal combustion engines since the motor vehicle overthrew the horse and buggy a century ago. None have thrived. Most have failed. Petroleum still dominates the transportation market in every country of the world. Initiatives to introduce alternative fuels have all fallen short. The successes that exist today are mostly limited to specific geographic markets where special circumstances have promoted alternative fuels. Oil sands production is booming in Canada, but that fuel is essentially identical to petroleum. Ethanol produced from corn sells well in the United States as a fuel additive, as does ethanol produced from sugar cane in Brazil, but both countries heavily subsidize its production and use. Compressed natural gas is used in limited amounts in Argentina and many other countries, including the United States, but the total number of vehicles is small and the fuel and vehicles receive substantial public subsidies. No alternative fuel in the market today is showing strong potential relative to the growth of gasoline and diesel vehicles.

Alternative fuels have failed largely because of unrealistic expectations and forecasts, which in turn led to unrealistic claims. These claims were usually not intentionally deceptive. Rather they reflected a poor state of knowledge of energy and environmental phenomena and policy. Three errors stand out: understated forecasts of oil supply, a false assumption that gasoline fuel quality could not be improved, and overstated environmental and economic benefits of alternative fuels. In reality, oil turned out, so far at least, to be cheap and abundant; gasoline and diesel fuel were reformulated to be cleaner; and new internal combustion engines are nearly zero emitting. These facts have breathed new life into conventional vehicle drivetrain designs and the petroleum fuels that power them.

The Case for Hydrogen

What do these lessons imply for hydrogen? First, hydrogen is unlikely to succeed on the basis of environmental and energy advantages—at least for the foreseeable future. Hydrogen is neither the easiest nor the cheapest way to gain large near and medium term societal benefits. Hybrid electric vehicles, cleaner combustion engines, and cleaner fuels will provide almost as much benefit on a per vehicle basis at a much lower cost. Because they can be commercialized much more quickly, they offer much more overall benefit sooner.

The case for hydrogen is compelling only over the long term, largely by creating a motor vehicle that is superior in performance to conventional vehicles, and also fundamentally different. Fuel cell vehicles are clearly superior environmentally to oil-burning vehicles. They also offer quiet operation, rapid acceleration from a standstill because of the torque characteristics of electric motors, and potentially low maintenance requirements. They provide most of the benefits of battery electric vehicles, without the short range and long recharge time. Moreover, fuel cell vehicles provide extra value to customers, as discussed in Chapter 3 by Ken Kurani *et al.* For example, they can provide remote electric power for use at construction sites, for recreational activities, and other applications. Fuel cell vehicles can even act as distributed electricity generators when parked at homes and offices.

Importantly, hydrogen fuel cell vehicles provide special attractions to automakers. By eliminating most mechanical and hydraulic subsystems, they provide greater design flexibility and the potential for using fewer vehicle platforms and therefore more efficient manufacturing approaches. As a result, the automotive industry, or at least an important slice of it, sees fuel cells as its inevitable and desired future. As noted by Jim Boyd in Chapter 10, automaker support was not evident in other movements to promote alternative fuels.

Cost and Availability Challenges

Fuel cells offer a superior consumer product when used in motor vehicles, but only if fuel cell costs become competitive and if hydrogen fuel can be made widely available at a reasonable cost. These two "ifs" remain unresolved and are central to the hydrogen debate. Fuel cell costs are on a steep downward slope, but they are still a factor of 10-20 too high. Huge amounts of engineering are still needed to improve manufacturability, assure long life and reliability, and withstand extreme temperatures. While some engineers, such as Geoffrey Ballard, believe that entirely new fuel cell architectures are needed to achieve the last 10-fold cost reduction, a handful of automotive companies seem convinced that they are on track to achieve the necessary cost reductions and performance enhancements, as Taiyo Kawai of Toyota states in Chapter 4.

The second "if," hydrogen availability, is perhaps the greater challenge. The problem is not production cost or sufficient resources. Hydrogen is already produced from natural gas and petroleum at costs similar to gasoline. Several chapter authors note that, with continuing R&D investment, the cost of producing hydrogen from a variety of fossil and renewable sources should not be much greater than producing gasoline. As petroleum fuel prices rise in the face of supply constraints, hydrogen production costs will become more competitive.

The key supply challenges, as articulated by Joan Ogden in Chapter 5, are as follows. First, because private investments will naturally gravitate toward fossil energy sources, government needs to accelerate R&D of production of renewable hydrogen so that it becomes a competitive option. There are many possible paths, and it is difficult at this time to know which will prevail. Second, carbon dioxide sequestration—a prerequisite if abundant coal in the United States, China, and elsewhere is to be used—faces uncertain public acceptance. Capturing and sequestering carbon dioxide, note Richard Doctor and John Molburg in Chapter 6, is not a showstopper on technological or cost grounds in many locations and situations.

The third supply-related challenge, and perhaps the most important of all, is the cost of delivering hydrogen from production sites to local fuel stations. Astoundingly, distribution and delivery of hydrogen to small hydrogen users now costs roughly five times as much as hydrogen production. Even at large fossil-based hydrogen production facilities under study, distribution and delivery costs are estimated to be as great as the cost of producing the hydrogen. Most problematic, especially during the low volume transition phase, is the "last mile" delivery of very low density hydrogen gas to the local fuel stations. Pipelines are the only plausible option for large volume hydrogen delivery, but they are too costly for small volume, short delivery shipments.

Today's existing natural gas and petroleum distribution systems may not be good models for future hydrogen distribution systems. If future hydrogen distribution systems mimic today's energy systems, the hydrogen economy might never happen because the construction costs would be too high and the undertaking could prove to be a logistical nightmare. An entirely new approach may be needed, one that serves both stationary and mobile users and small as well as large hydrogen production facilities. It must also access a wide variety of energy feedstocks, incorporate carbon dioxide capture and sequestration, and be geographically diverse. Such a system won't happen by itself, nor through the leadership of today's oil and gas companies. New ideas and approaches are needed to launch a new hydrogen industry in a precommercial environment where long term investment capital is scarce. Hydrogen entrepreneurship, discussed in the chapters by David Bodde (Chapter 7) and Chip Schroeder (Chapter 8), will undoubtedly focus on distributed generation, with production near or at the end-use site. The hydrogen economy will initially be based on distributed generation of hydrogen once excess capacity at existing industrial hydrogen production facilities is brought to use. This solution, which may prevail for a long time, could include the placement of small hydrogen refueling appliances at residences, a concept now being investigated by Honda and General Motors as well as a number of small hydrogen start-up companies. Other innovative solutions would be needed, especially during the early phases when consumption is small and dispersed.

A Precarious Situation

The challenges facing hydrogen remain daunting. But interest in hydrogen remains strong, not only because of the potential energy and environmental benefits and automaker enthusiasm, but also because of the absence of a more compelling long term option. The only other serious long term proposal is grid-supplied electricity. But that requires further improvements in onboard electricity storage devices, or massive investments in "third rail" electricity infrastructure, including substantial added cost for vehicles. Other candidates include biomass, but supply would not be adequate in most regions of the world, and fossil-based synfuels, which would emit huge amounts of carbon into the atmosphere and require massive networks of pipes to deliver carbon dioxide to hopefully secure sequestration sites.

A key aspect of the hydrogen economy, currently with unknown consequences, is hydrogen's inclusivity. It can be produced from virtually any energy feedstock, including coal, nuclear, natural gas, biomass, wind, and solar. This characteristic, along with hydrogen's environmental benefits, results in no natural political or economic enemies. Even oil companies, in actuality massive energy companies, are ambivalent. They are well prepared to supply any liquid or gaseous fuel consumers might desire, though they prefer a slow transition that allows them to protect sunk investments. Most, for instance, prefer that initial fuel cell vehicles carry reformers to convert gasoline into hydrogen. They have been disappointed to see all major car companies now focused strictly on delivered hydrogen.

A critical difference between automaker and oil company business realities is that some automakers will likely see benefits from being first to market. Oil companies see little advantage in being a pioneer. They see a prolonged financial drain. They are unlikely to be early investors in a rapid buildup of hydrogen fuel stations. That financial burden will fall on government.

Much is at stake, and it is not surprising that skepticism is already coming from many quarters. Academics question near term environmental benefits, as they have in recent *Science* articles, and activists and environmental groups such as Jeremy Rifkin and Sierra Club question the social, environmental, and political implications of what they call "black" hydrogen made from coal or using nuclear energy. Others say hydrogen is the wrong horse to pick, as Paul MacCready does in Chapter 16, when he argues that improved battery technology will trump hydrogen and fuel cell vehicles.

Many argue that the hydrogen transition is premature, as does John DeCicco in Chapter 15.

The mounting hydrogen debate is being sucked into the larger debate over the Bush administration's environmental record. The environmental community fears a pact between the administration and car and oil companies to "greenwash" industry efforts and camouflage eviscerated and stalled regulations. Where some see a progressive long term strategy, others see a conspiracy to promote conventional technologies under the guise of transforming the energy system into a hydrogen economy. A backlash is also building against what many see as hydrogen hype.

It is clear, however, that hydrogen is different from all previous alternative fuel experiences, largely related to the business attractions of fuel cells. Therein lies the source of the surprisingly strong interest in hydrogen. Because automakers are the interest group that sees the greatest payoff, hydrogen's future appears to be tightly linked to automaker commitments to move fuel cells from the laboratory into the marketplace. The key question is whether and when they will ratchet up current investments of perhaps $150 million per year in the case of the more aggressive automakers to the much larger sums needed to tool factories and launch commercial products. Without automaker leadership, the transition will be slow, building on small entrepreneurial investments in niche opportunities—such as fuel cells in forklifts and other off-road industrial equipment, hydrogen blends in natural gas buses, and innovative low-cost delivery of hydrogen to small users. It is uncertain to what extent innovative small hydrogen and fuel cell companies are key to the transition—in a political as well as commercial sense—and what role government should play in supporting those small companies. Another unknown is whether the involvement of major companies in the hydrogen transformation can exist without stifling the entry of new hydrogen and fuel cell companies that might be needed to make the hydrogen transition a success.

In the end, the hydrogen situation is precarious. Beyond a few car companies, government agencies, and scattering of entrepreneurs, academics, and environmental advocates, support for hydrogen is thin. While many rail against the hydrogen hype, the greater concern perhaps should be the fragile support for hydrogen. Politics aside, many applaud government efforts to start down a path toward a sustainable future. Policy understandably must remain focused on the near term, but those near term policies should also be sending signals to businesses and customers that lead to socially beneficial long term outcomes.

Despite a wide diversity of opinion that exists in energy policy circles today about what policies are needed to hitch the present energy wagon to the hydrogen star, a near consensus has emerged in recent years that the hydrogen economy is a great hope to meet the world's future energy needs. This book provides the foundation of knowledge and the diversity of perspectives that will lead to more informed policy and business debates on how and when to take the next steps in the hydrogen transition.

APPENDIX A

About the Editors and Authors

About the Editors

Daniel Sperling is Professor of Civil Engineering and Environmental Science and Policy, and founding Director of the Institute of Transportation Studies (ITS-Davis) at the University of California, Davis. Dr. Sperling is recognized as a leading international expert on transportation technology assessment, energy and environmental aspects of transportation, and transportation policy. In the past 20 years, he has authored or coauthored over 160 technical papers and 8 books. Dr. Sperling is associate editor of *Transportation Research D (Environment)* and a current or recent editorial board member of five other scholarly journals. He is a recent member of U.S. National Academy of Sciences committees on Highway Finance (2003–2004), Hydrogen Production and Use (2002–2003), Personal Transport in China (2000–2002), Transportation Environmental Cooperative Research Program Advisory Board (1999–2001), Biomass Fuels R&D (1999), Enabling Transportation Technology R&D (1998), Transportation and a Sustainable Environment (1995–1997), Transportation Options for Megacities (1994), and Liquid Fuel Options (1989-1990). He is founding chair and emeritus member of the Alternative Transportation Fuels Committee of the U.S. Transportation Research Board. Dr. Sperling consults for international automotive and energy companies, major environmental groups, and several national governments. He has testified numerous times to the U.S. Congress and various government agencies, and provided keynote presentations and invited talks in recent years at international conferences in Asia, Europe, and North America.

James S. Cannon is an internationally recognized researcher specializing in energy development, environmental protection, and related public policy issues. He is president of Energy Futures, Inc., which he founded in 1979. Among its activities, Energy Futures publishes the quarterly international journal *The Clean Fuels and Electric Vehicles Report* and the bimonthly newsletter, *Hybrid Vehicles*. Mr. Cannon has written several books on alternative transportation fuels, including *The Drive for Clean Air* (1989); *Paving the Way to Natural Gas Vehicles* (1993); and *Harnessing Hydrogen: The Key to Sustainable Transportation* (1995). He has also written a number of in-depth studies and reports analyzing aspects of the transition of the global transportation energy system from oil to renewably produced hydrogen. Over the past decade, his research into alternative transportation fuels has taken him to over 20 countries on 5 continents. Mr. Cannon currently serves as senior fellow at INFORM, Inc., a nonprofit environmental and public health research organization. Previously, he had a 7-year professional association with the U.S. Office of Technology Assessment, and for 8 years was an energy policy analyst for the Energy, Minerals, and Natural Resources Department of the State of New Mexico. Mr. Cannon holds an AB in chemistry from Princeton University and an MS in biochemistry from the University of Pennsylvania. He lives with his family in Boulder, Colorado.

About the Authors

Loren K. Beard is senior manager for Environmental and Energy Planning at DaimlerChrysler's U.S. headquarters in Auburn Hills, Michigan. He is responsible for developing policy and planning strategy for fuel economy, fuel quality, global climate change, and stationary source energy usage. Prior to this assignment he worked as senior manager—fuels, in DaimlerChrysler Materials Engineering. Dr. Beard came to the former Chrysler Corporation in 1993 after 6 years with Amoco Oil R&D in Naperville, Illinois (1987–1993), and 6 years at BASF R&D in Wyandotte, Michigan (1981–1987). He holds a BS in chemistry from Oakland University in Rochester, Michigan, and PhD in inorganic chemistry from The Pennsylvania State University.

David L. Bodde serves as the Charles N. Kimball Chair in Technology and Innovation at the University of Missouri in Kansas City. He was a member of the National Research Council committee studying the hydrogen economy. Dr. Bodde serves on the Board of Directors of several energy and technology companies, and advises many others on technology based strategy. He has a wide range of executive experience in both government and the private sector, including the Midwest Research Institute, the Congressional Budget Office, the Department of Energy, and TRW. He served in the Army in Vietnam.

James D. Boyd was appointed to the California Energy Commission in 2002 by Governor Gray Davis to serve a 5-year term. Commissioner Boyd presides over the Energy Commission's Transportation Committee and oversees

Climate Change and International Export Programs. He also serves on the Electricity and Natural Gas and Renewables Committees and chairs the state's Joint Agency Climate Change Team and the state's Natural Gas Working Group. He is chairman of the committee overseeing the preparation of the Energy Commission's *2003 Integrated Energy Policy Report.* He serves as the state's liaison to the Nuclear Regulatory Commission. Commissioner Boyd is co-chairman of the Border Governors' Conference Energy Worktable and represents the Energy Commission on the Steering Teams of the California Fuel Cell Partnership and the California Fuel Cell Collaborative. He serves on the Board of Directors of WESTART/ CALSTART and on the Board of Advisors at the ITS-Davis. A California native, Commissioner Boyd received his BS in business administration from the University of California, Berkeley.

Steve Chalk is the director of the Office of Hydrogen, Fuel Cells and Infrastructure Technologies at the Department of Energy (DOE) and is the primary point of contact for hydrogen-related activities in the department for the president's Hydrogen Fuel Initiative. Prior to this, Mr. Chalk managed the Energy Conversion Team in DOE's Office of Advanced Automotive Technologies, where he was responsible for fuel cells, compression-ignition, direct-injection engines, advanced fuels, and materials. Mr. Chalk holds a BS in chemical engineering from the University of Maryland and a MS in mechanical engineering from George Washington University.

Christopher D. Congleton is a third-year PhD student in Transportation Technology and Policy at UC Davis. His current research explores the introduction of battery electric vehicles in California, their market potential, and the potential effects of such vehicles on household activity spaces and the consequences of such changes on urban development and neighborhood design. He produces educational films on such topics as air quality, global warming, oil dependence, and electric-drive vehicles. Mr. Congleton holds a BA in culture and technology from the University of California at Santa Cruz. During his undergraduate work he studied and lectured on publicly available technologies and practices to improve energy systems and expand lifestyle choices. He is a member of the Institute of Transportation Engineers and the American Society of Cybernetics.

John M. DeCicco is a senior fellow who specializes in automotive strategies at Environmental Defense, a national environmental group. His work entails technology assessment and policy analysis of ways to improve efficiency and reduce emissions of cars and light trucks. He has published extensively on the subject, with recent studies addressing options for improving the fuel economy of gasoline powered vehicles, including conventional and hybrid power trains and mass reduction; prospects for fuel cell vehicles; and market characterizations of automotive sector oil demand and CO_2 emissions.

He also has an interest in developing regulatory, market oriented, and consumer educational strategies to foster progressive change in the auto market. Dr. DeCicco is active in the Society of Automotive Engineers as well as the National Research Council's Transportation Research Board, for which he chaired the Energy Committee from 1996–2000. He received his doctorate in mechanical engineering from Princeton University in 1988.

Reid R. Heffner is currently a graduate student in the Transportation Technology and Policy Graduate Group at the University of California, Davis and a graduate student researcher at the ITS-Davis. Before starting graduate studies at UC Davis, Mr. Heffner worked in the development and marketing of enterprise software applications. Most recently, he was a Senior Product Manager with a supply-chain software company, where he worked with numerous clients in the motor vehicle industry. He has a BA from Colgate University, and a MBA from Georgetown University.

Lauren Inouye works for Sentech Inc. in support of the DOE Hydrogen Program. She graduated from Stanford University in 2002 with a degree in international relations with a focus on energy and environmental policy. Following a fellowship with the Clean Energy Group at Winrock International, an international development group, Ms. Inouye joined Sentech assisting the DOE Hydrogen Program on communications and international work.

Kenneth S. Kurani is a member of the professional research staff at ITS-Davis. Working as part of a multidisciplinary team, he develops and applies methods to evaluate user responses to new transportation and information technologies. This research is conducted within a household-activity-based approach to travel demand analysis and interactive stated preference and reflexive survey methodologies. His research explores how citizen-consumers can use new technologies to shape both their own lives as well as efforts to market transportation and communication networks according to their collective benefits such as energy efficiency, air quality, safety, and social equity. His ongoing research includes household response to electric, hybrid, and fuel cell vehicles and consumer-citizen valuation of automotive fuel efficiency. Dr. Kurani holds a PhD in civil and environmental engineering from UC Davis.

Paul B. MacCready is chairman and founder of AeroVironment, Inc., a pioneering technology developer since 1971. In the 1950s, Dr. MacCready led a team of researchers investigating the mechanisms and effectiveness of cloud seeding. He has been considered "the father of human powered flight" since the late 1970s, when his Gossamer Condor and Gossamer Albatross captured the first two Kremer Prizes, the largest cash awards in aviation. His continued efforts with the teams at AeroVironment fit the themes of energy efficiency, and conservation. The Solar Challenger carried its pilot 263 kilometers from Paris to England solely on solar energy. His teams also created the GM Sunraycer that won the first solar car race across Australia. Two years later, Dr. MacCready and his team built the General Motors Impact battery electric demonstrator car, the predecessor to the GM production EV-1. In 2001, he built the solar powered Helios aircraft that flew to 29,523 meters, setting a world altitude record. Dr. MacCready received a doctorate in aeronautics from the California Institute of Technology in 1952.

Barry D. McNutt was a senior policy analyst in the Office of Policy at the U.S. Department of Energy in Washington, D.C. until his death in November 2003. Mr. McNutt had primary responsibilities in the analysis and policy development for transportation efficiency, alternative fuels, clean fuels, and reformulated gasoline issues. Mr. McNutt was the lead analyst and technical representative on clean fuel regulatory and refinery issues and chaired the Coordinating Subcommittee and the Producibility Workgroup of the National Petroleum Council's Refining and Product Deliverability Study. Mr. McNutt was a member of EPA's Blue Ribbon Panel on Oxygenates and the Phase II RFG Task Group and he chaired other NPC studies on refining and petroleum product inventory. Mr. McNutt received a BA and BS in mechanical engineering from Rutgers University. Prior to his work for DOE, he was with the Federal Energy Administration and the Environmental Protection Agency. He authored numerous technical and policy papers and presentations on transportation energy use, and fuels and fuel economy, and was an active participant in many Asilomar transportation conferences. This book is dedicated to the memory of Mr. McNutt.

Joan Ogden is Associate Professor of Environmental Science and Policy at the University of California, Davis and an associate energy policy analyst and co-director of the Hydrogen Pathway Program at the ITS-Davis. Her primary research interest is technical and economic assessment of new energy technologies, especially in the areas of alternative fuels, fuel cells, renewable energy, and energy conservation. Her recent work centers on the use of hydrogen as an energy carrier, particularly hydrogen infrastructure strategies, and applications of fuel cell technology in transportation and stationary power production. She has written extensively on energy topics, including one book, eight book chapters, and numerous peer reviewed journal articles and conference presentations. Dr. Ogden headed the systems integration team for the U.S. DOE National Hydrogen Roadmap in 2002, and is active in the H2A, a group of hydrogen analysts convened by the Department of Energy to develop a consistent framework for analyzing hydrogen systems. Dr. Ogden received a PhD in theoretical plasma physics from the University of Maryland in 1977. She was a research scientist at Princeton University's Center for Energy and Environmental Studies from 1985 to 2003. She joined the faculty of UC Davis in September 2003.

Bernard Robertson is the Senior Vice President—Engineering Technologies & Regulatory Affairs at DaimlerChrysler Corporation. He was elected an officer of Chrysler Corporation in February 1992. He was appointed senior vice president coincident with the merger of Chrysler Corporation and Daimler-Benz AG in November 1998, and was named Senior Vice President—Engineering Technologies and Regulatory Affairs in January 2001. Mr. Robertson is responsible for the Liberty and Technical Affairs Research group, Advanced Technology Management, FreedomCAR, Dodge Truck HEV, battery electric and military vehicle development. In addition, he is responsible for regulatory analysis and compliance, including safety and emissions, both mobile and stationary sources. Mr. Robertson is a General Aviation pilot and owns two aircraft. He holds a MBA. from Michigan State University, 1976 and a masters degree, automotive engineering, Chrysler Institute, 1967, and a masters degree, mechanical sciences, Cambridge University, England 1967. He holds both British and American citizenship.

David E. Rodgers is senior advisor to the Deputy Assistant Secretary for Technology Development within the Office of Energy Efficiency and Renewable Energy at the U.S. DOE in Washington, D.C. Mr. Rodgers was previously the director of the Office of Technology Utilization within the Office of Transportation Technologies, also at the DOE. During his career, Mr. Rodgers has had budgeting, management, and planning responsibilities for the alternative fuel and advanced vehicle provisions of the Energy Policy Act of 1992, including introduction of alternative fuel vehicles into the federal fleet, the Clean Cities program, public information, state grants, and regulatory development. Mr. Rodgers received a BS in chemical engineering and a BS in computer science from Washington University in St. Louis, and a masters in public management from the University of Maryland. In the private sector, Mr. Rodgers has experience in the chemical, petroleum, and computer industries.

Jonathan Rubin is interim director of the Margaret Chase Smith Center for Public Policy and Associate Professor in the Department of Resource Economics & Policy at the University of Maine. He received his doctorate in agricultural economics from the University of California, Davis, in 1993. He also holds a master's degree in economics from the University of Washington. He is a member of the Transportation Research Board's, Alternative Fuels and Transportation Energy Committees. Dr. Rubin's research focuses on market based solutions to attain environmental goals. Working with colleagues at Oak Ridge National Laboratory, he studies the economics of hybrid, fuel cell, and other alternative transportation vehicles and fuels in the United States. Dr. Rubin is a coauthor of the Transitional Alternative Fuels and Vehicles model that numerically simulates transitional barriers to adopting new vehicle and fuel technologies. Other recent publications have investigated the potential economic and social impacts involved in the trading of criteria and greenhouse gas pollution credits.

David Sanborn Scott was born in Quebec City and lived his preschool years in the jungles of Belize and the forests of northern Quebec. He obtained his undergraduate degree in mechanical engineering from Queen's University in 1959. After several years in industry, he received a doctorate in mechanical engineering and astronautical sciences from Northwestern University. In 1966, Dr. Scott joined the Department of Mechanical Engineering at the University of Toronto and became departmental chair in 1976. He founded the University of Toronto's Institute of Hydrogen Systems—and then chaired the Canadian Advisory Group on Hydrogen Opportunities that produced the 1987 report *Hydrogen: National Mission for Canada*. In 1989, David joined the University of Victoria, where he founded its Institute for Integrated Energy Systems (IESVic). IESVic is developing fuel cell systems and cryofuel liquefaction, and performing energy systems analysis, all with industrial and/or government collaboration. Dr. Scott is Vice President for the Americas for the International Association for Hydrogen Energy and is completing a book *Smelling Land: An Odyssey towards a Cleaner, Richer Hydrogen Age* scheduled for release in late 2004.

Thomas S. Turrentine is an anthropologist at ITS-Davis. His research interests include the role of travel and movement in the shaping of culture. He focuses on automobile based lifestyles, developing methods and theories to explore potential responses to new technologies and policies. Past research has included studies of consumers and alternative fueled vehicles, natural gas vehicles, station cars, neighborhood electric vehicles, and traveler information systems. Recent research includes markets for fuel cell vehicles, consumers and fuel economy, road ecology, and social marketing as a solution to environmental dilemmas in transportation markets. Dr. Turrentine holds a PhD in anthropology from UC Davis.

Barend C.W. van Engelenburg has been employed by the Ministry of Environment in The Netherlands since 1997. His main task is to advise the minister on long term technology issues related to climate change and sustainable transport. The focus of his work is primarily international in scope. He has extensive experience in stimulating the development of alternative fuels. He is the commissioner of the Dutch GAVE program, an important program in this area. Since 1997, he has also been a member of the Executive Committee of the IEA Greenhouse Gas R&D Program. Last year, he became the Dutch IPCC representative for Working Group III and is further involved in a couple of European Union activities in the area of sustainable transport and alternative fuels. His assignment at the Energy Modeling Forum at Stanford University in the summer of 2003 was part of his international activities. Before 1997, he was employed at Utrecht University for 7 years, mainly involved in energy and lifecycle analysis of energy systems but also in some ad hoc work such as analyzing CO_2 capture and storage, solar cells, and indirect household energy use.

APPENDIX B

Asilomar Attendee List 2003

First & Last Name	*Organization*
Don Anair	Union of Concerned Scientists
George Anderson	Natural Resources Canada
Kristin Andrichik	Shell Hydrogen
Patrick Achard	Amines-Centre d'Energetique
Juan Argueta	Southern California Edison
Michael Ball	Transport Canada
Geoffrey Ballard	General Hydrogen
J.P. Batmale	Great Valley Center
Shannon Baxter	California Air Resources Board
Richard Bechtold	QSS Group, Inc.
Louise Bedsworth	Union of Concerned Scientists
Andrew Bermingham	Hydrogen Works LLC
Analisa Bevan	California Air Resources Board
Dave Bodde	University of Missouri
John Boesel	WestStart - CalStart
Bill Boyce	Sacramento Municipal Utility District
Jim Boyd	California Energy Commission
John Bradley	Edwards & Kelcey
Harold Brazil	Metropolitan Transportation Commission
William Breed	U.S. Department of Energy
C.J. Broderick	University of California, Davis
Mary Jean Burer	Natural Resources Defense Council
Larry Burns	General Motors
John Cabaniss	Association of International Automobile Manufacturers
Tom Cackette	California Air Resources Board
Joseph Calhoun	California Air Resources Board
Kateri Callahan	Electric Drive Transportation Association
Bill Canella	ChevronTexaco
James Cannon	Energy Futures
Leslie Caplan	Blue Water Group
Steve Chalk	U.S. Department of Energy

Continued

Asilomar Attendee List 2003—cont'd

First & Last Name	*Organization*
Robert Chapman	The RAND Corporation
Woodrow Clark	Governor's Office of Planning and Research
Jeff Cooke	ConocoPhillips
James Corbett	University of Delaware
Michael Cummings	PEW Center on Global Climate Change
Achim Dahlen	Ministry of Transportation and Spatial Planning, Germany
Greg Dana	Alliance of Automobile Manufacturers
Danielle Deane	Hewlett Foundation
John DeCicco	Environmental Defense
Mark Delucchi	UC Davis
Todd Dipoala	Kirsch Foundation
Clarence Ditlow	Center for Auto Safety
Gary Dixon	South Coast Air Quality Management District
Richard Doctor	Argonne National Laboratory
Bill Dougherty	Tellus Institute
Thomas Drennen	Sandia National Laboratories
Bill Drumheller	International Council for Local Environmental Initiatives
K.G. Duleep	Energy and Environmental Analysis
Catherine Dunwoody	California Fuel Cell Partnership
Harry Dwyer	UC Davis
Michael Eaves	California Natural Gas Vehicle Coalition
Anthony Eggert	UC Davis
Daniel Emmett	Environment Now Foundation
Bob Epstein	Environmental Entrepreneur (NRDC)
Tony Estrada, Jr.	Pacific Gas & Electric
Alex Farrell	UC Berkeley
Barbara Finamore	Natural Resources Defense Council
Todd Fronckowiack	Ford Motor Company
Ichiro Fujimoto	Takushoku University
Lew Fulton	International Energy Agency
David Garman	U.S. DOE
Devinder Gerewal	California Air Resources Board
John German	American Honda Motor Corporation
Lorne Gettel	IMPCO Technologies, Inc.
Norma Glover	CNGVP
David Greene	Oak Ridge National Laboratory
Charles Griffith	Ecology Center
Edmund Grostick	Oak Ridge National Laboratory
Chris Grundler	U.S. EPA
Don Hardesty	Sandia National Laboratories
Andrew Hargadon	UC Davis
David Hart	Imperial College Centre Environmental

Asilomar Attendee List 2003—cont'd

First & Last Name	*Organization*
Karl Hauer	xcellvision
David Hawkins	Natural Resources Defense Council
Dave Hermance	Toyota Technical Center
John Heywood	Massachusetts Institute of Technology
Michael Hickner	UC Davis
Toshio Hirota	Nissan Technical Center North America
David Hitchcock	Houston Advanced Research Center
John Hutchison	Ministry of the Environment, Canada
Lauren Inonye	Sentech, Inc.
Mike Jackson	TIAX
Brian Johnston	Nissan Technical Center North America
Larry Johnson	Argonne National Laboratory
Jack Johnston	ExxonMobil Research and Engineering
Mike Jones	BP Gas Marketing, Ltd.
Takehiko Kato	Interlink Corporation
Taiyo Kawai	Toyota Motor Corporation
Jay Keller	Sandia National Laboratories
Paul Khanna	Natural Resources Canada
Ben Knight	Honda R&D Americas
Fumiyaki Kobayashi	Toyota Motor Sales, U.S.A.
Axel Koenig	Volkswagen AG
Wilfrid Kohl	Johns Hopkins University
Drew Kojak	National Commission on Energy
Ken Koyama	California Energy Commission
Joe Krovoza	UC Davis
Ken Kurani	UC Davis
Robert P. Larsen	Argonne National Laboratory
Eric Larsen	DaimlerChrysler Research & Technology
Bob Larson	U.S. EPA, Office of Transportation & Air Quality
Michael Lawrence	Jack Faucett Industries
Paul Leiby	Cak Ridge National Laboratory
Lewison Lem	California State Automobile Association
Buford Lewis	ExxonMobil Corporation
Amy Lilly	American Honda
Tim Lipman	UC Davis, UC Berkeley
Alan Lloyd	California Air Resources Board
Mike Love	Toyota Motor Sales, U.S.A.
Jeffra Lyczko	DaimlerChrysler Society & Technology Research Center
Elisa Lynch	Blue Water Group
Paul MacCready	AeroVironment, Inc.
Jerry Mader	Center for Automotive Research
Margaret Mann	National Renewable Energy Laboratory

Continued

Asilomar Attendee List 2003—cont'd

First & Last Name	*Organization*
Jason Mark	Union of Concerned Scientists
Dave McCarthy	Air Products & Chemicals
Barry McNutt	U.S. Department of Energy
Mark Melaina	University of Michigan
Scott Miller	Synovate Motoresearch
Kevin Mills	Environmental Defense
Marianne Mintz	Argonne National Laboratory
Patricia Monahan	Union of Concerned Scientists
Bob Moore	UC Davis
Gene Nemanich	ChevronTexaco Technology Ventures
Joan Ogden	UC Davis
Neil Otto	Independent
Ben Ovshinsky	ECD Ovonics
Bob Pahl	ConocoPhillips
Michel Parent	INRIA (France)
George Parks	ConocoPhillips
Geoff Partain	Toyota Motor Sales, U.S.A.
Mark Paster	U.S. Department of Energy
Charlotte Pera	Energy Foundation
Lori Priday	Nissan North America
Jim Ragland	ARAMCO Services Company
Chella Rajan	Tellus Institute
Sharima Rasanayagam	British Consulate General
Deborah Redman	UC Irvine
Peter Reilly-Roe	Natural Resources Canada
Bill Reinert	Toyota Motor Sales, U.S.A.
Giorgio Rizzoni	Ohio State University
Bernard Robertson	DaimlerChrysler
David Rodgers	U.S. Department of Energy
Bob Rose	U.S. Fuel Cell Council
Philip Ross	Lawrence Berkeley National Laboratory
Jonathan Rubin	University of Maine
Richard Rykowski	U.S. Environmental Protection Agency
Mattia Sattone	ChevronTexaco Technology Ventures
Lee Schipper	World Resources Institute
Chip Schroeder	Proton Energy Systems
Dennis Schuetzle	The Renewable Energy Institute
David Scott	University of Victoria
Jim Seaba	ConocoPhillips
Phil Sharp	Harvard University
Harry Sigworth	ChevronTexaco Energy Research & Technology
Margaret Singh	Argonne National Laboratory
Jan Skogstrom	Independent Filmmaker

Asilomar Attendee List 2003—cont'd

First & Last Name	*Organization*
Ray Smith	Lawrence Livermore National Laboratory
Dan Sperling	UC Davis
Elise Steiner	National Renewable Energy Laboratory
Brian Stokes	Pacific Gas & Electric
Detlef Stolten	Research Center Juelich, Germany
Terry Surles	California Energy Commission
George Sverdrup	NREL
Fujio Takimoto	Subaru Technical Research Center
Ruth Talbot	Natural Resources Canada
Bob Thompson	Tahoe Systems
Greg Thompson	West Virginia University
Tom Turrentine	UC Davis
Jim Uihlen	BP
Zev Ungar	Renault
Stefan Unnasch	TIAX
Yoichi Uraki	Nissan Technical Center North America
Baren van Engelenburg	Ministry of the Environment, The Netherlands
Arnold Van Zyl	EUCAR (ACEA)
Francis Vanek	Cornell University
Michael Walsh	International Consultants
W. Scott Wayne	West Virginia University
Hank Wedaa	Valley Environmental Associates
Tom White	U.S. Department of Energy
Carol Whiteside	Great Valley Center
Brett Williams	UC Davis
Roy Wilson	South Coast Air Quality Management District
Steven Winkelman	Center For Clean Air Policy
Catherine Witherspoon	California Air Resources Board
Chris Yang	UC Davis
Aki Yasuoka	American Honda Motor Company
Chicheng Yeh	TECO
John Ziagos	Lawrence Berkeley National Laboratory
Karl Heinz Ziwica	BMW North America

INDEX

A

AFCG (Alternative Fuels Contact Group), 157
AFV penetration in markets, 178
AFVs (alternative fuel vehicles), 15, 167, 181–190, 192
  analysis and experience of, 191–212
  battery powered electric vehicles (BEVs), 185–186
  Chrysler's experience in deployment of, 182
  compressed natural gas (CNG), 183
  E85, 184–185
  holistic approach to national energy policy, 181–182
  hydrogen powered fuel cell vehicles, 188–189
  light duty diesel vehicles, 186–188
  M85, 183–184
  percent of sales of, 169
  small market segment and buying, 202–204
Agents of change, entrepreneurs as, 112–113
Air, global mass of vertebrate life in, 230
Air pollutant emissions, well-to-wheels, 81
Air Resources Board (ARB), 149
Alcohol fuel vehicles, 171
Alternative fuel programs, 166–167
Alternative fuels
  key policy initiatives and, 167–168
  social attributes of new, 176
Alternative Fuels Contact Group (AFCG), 157
Alternative fuels experience, 15 years of, 165–179
  alternative fuels experience is valuable, 165–166
  effort of United States, 168–172
Alternative fuels experience *(Continued)*
  hydrogen transition and alternative fuel programs, 166–167
  key policy initiatives and alternative fuels, 167–168
  lessons learned, 172–178
Alternative Motor Fuels Act of 1988 (AMFA), 167, 192, 215
Alternatives
  competing automotive, 51–52
  hydrogen as battery, 127–129
  to hydrogen today, 17–18
AMFA (Alternative Motor Fuels Act of 1988), 167, 192, 215
Analyses
  energy cycle, 97–99
  transitional, 193–195
ARB (Air Resources Board), 149
Arbitrage
  electrolysis as energy, 131–133
  energy, 132
Architecture of energy system, 25–27
Asilomar Conference, 7–8
Assembled products, innovation in, 110–111
Autos; *See also* Cars
Auto and oil industries, coordination between, 176
Automobility and communications, 46–47
Automotive alternatives, competing, 51–52
Availability challenges, cost and, 236–238

B

Barriers, long-run, 204–206
Batteries, lithium, 231–233
Battery alternative, hydrogen as, 127–129

BEVs (battery electric vehicles), 185–186
BEVs (battery electric vehicles), case for, 217, 227–233
  challenge of sustainability, 228–231
  global mass of vertebrate life on land and in the air, 230
  lithium batteries as solution, 231–233
  solar energy supply, 229
BTL (biomass-to-liquids), 188
BTUs (British thermal units), 123
Business model, 115–117

C

CAFE (corporate average fuel economy), 168, 185, 192, 221
California Energy Commission (CEC), 13, 150
Capture and sequestration, CO_2, 93–104
CARB (California Air Resources Board), 13, 149
Carbon taxes and CO_2 sequestration, 102–103
Cars, Toyota's pathways for eco, 60
CEC (California Energy Commission), 13, 150
CERS (cryogen energy recovery systems), 29
Chains, fuel, 159
Challenges, cost and availability, 236–238
Changes
  entrepreneurs as agents of, 112–113
  rationales for, 216–217
Characteristics, hybrid system, 64
Chicken or egg problem, 213–226
Choices, 202
  evaluating strategic, 217–223
Chrysler's experience in deployment of AFV's, 182
Clean Urban Transport for Europe (CUTE), 162
CNG (compressed natural gas), 166, 183
CO_2 (carbon dioxide), 10, 59, 73, 149, 182
  capture and sequestration, 93–104
  transportation, 99–101
CO_2 emissions, well-to-wheel, 69
CO_2 sequestration, 101–102
  carbon taxes and, 102–103
Coal, clean hydrogen from, 93–104
Coal-derived hydrogen, 95–97
Coal-to-hydrogen, energy cycle analysis of, 97–99
Commercialization, overcoming barriers to, 67–71
Commitments
  history of EU, 156–158
  history of U.S. hydrogen, 155–156
Communications, automobility and, 46–47
Compressed natural gas (CNG), 166, 183
Conference, Asilomar, 7–8
Consumer markets, 174
Consumer reticence, 175
Consumers, 176
Continental energy distribution, 29–30
Conventional technology and forestalling transition, 206
Coordination, 202
Corporate average fuel economy (CAFE), 168, 185, 192, 221
Cost and availability challenges, 236–238
Cost reduction, technological, 198–199
Costs
  hydrogen production, 70
  of limited diversity, 197–198
  per model, 197
  of retail fuel availability, 196
Cryogen energy recovery systems (CERS), 29
CUTE (Clean Urban Transport for Europe), 162
Cycle analysis, energy, 97–99
Cycles, technology validation, 142

D

Demand, matching supply and, 84–86
Design, FCV and fuel infrastructure performance and, 52
Development, market, 175
Diesel vehicles, light duty, 186–188
Distribution, continental energy, 29–30
Diversity
  accounting for vehicle model, 197–198
  effective costs of limited, 197–198
DOE (Department of Energy), 7, 106, 135, 165, 192
DOE pathway to hydrogen through 2015, 141
DOE's R&D strategy, 139–142
Driving range and nontravel use of energy, 52–54

E

E85, 184–185
EC (European Commission), 14
ECERS (enhanced cryogen recovery systems), 29
Eco car, Toyota's pathways for, 60
Ecological City Transport System (ECTOS), 162
Economies of scale, 197
Economy, international partnership for hydrogen, 144–145
ECTOS (Ecological City Transport System), 162
EEA (Energy and Environmental Analysis), 199
Efficiency, well-to-wheel, 65
EIA (Energy Information Administration), 195
Electricity, 40–41
  hydrogen and, 23–24
  mobile, 47–48
Electricity generators, solar thermal hydrogen and, 130
Electrolysis
  efficiencies of hydrogen, 124
  as energy arbitrage, 131–133
  PEM, 123–126
  technology pathways, 122
Electrolysis, hydrogen from, 121–133
  electrolysis as energy arbitrage, 131–133
  fuel from renewable resources, 129–131
  hydrogen as battery alternative, 127–129
  PEM electrolysis, 123–126
  power quality, 127–129
  Proton Energy Systems, 121–123
  today's industrial hydrogen products, 126–127
Electrolytic hydrogen, energy arbitrage from, 132
Emissions
  lower vehicle, 170
  well-to-wheel CO_2, 69
  well-to-wheel GHG, 81
  well-to-wheels air pollutant, 81
Enabling technologies, 114
End game, hydrogen, 8–9
Energy arbitrage, electrolysis as, 131–133
Energy arbitrage from electrolytic hydrogen, 132
Energy cycle analysis of coal-to-hydrogen, 97–99
Energy distribution, continental, 29–30
Energy, driving range and nontravel use of, 52–54
Energy Information Administration (EIA), 195
Energy infrastructure, 39–41
  electricity, 40–41
  gasoline, 39–40
Energy policies, implications for, 117–118
Energy Policy Act of 1992 (EPACT), 165, 192, 215
Energy policy, holistic approach to national, 181–182
Energy security, 221–222
  and environment, 4–7
Energy system
  architecture of, 25–27
  existing, 87–88
Enhanced cryogen recovery systems (ECERS), 29
Entrepreneurs as agents of change, 112–113
Entrepreneurship, hydrogen, 11–13
Entrepreneurship in hydrogen transition, 105–119
  business model, 115–117
  current policy focus, 105–106
  enabling technologies, 114
  entrepreneurs as agents of change, 112–113
  how problem, 106–107
  implications for energy policies, 117–118
  innovation in assembled products, 110–111
  innovation in nonassembled products, 111–112
  innovation perspective, 108–110
  service concept for technology, 107–108
  starting in niche markets, 113–114
  transportation, 105–106
Environment, energy security and, 4–7
EPA (Environmental Protection Agency), 172
EPACT (Energy Policy Act of 1992), 165, 192, 215
EPACT goals, percent of sales of AFVs and, 169
Ethanol, 171
EU commitment, history of, 156–158

EU (European Union), 13
European perspective of US hydrogen activities, 155–164
  comparing both approaches, 158–162
  history of EU commitment, 156–158
  history of U.S. hydrogen commitment, 155–156
European Union (EU), 13

F

Factors, market, 223
FCHV fuel cell vehicle, Toyota, 61–62
FCHVs (fuel cell hybrid vehicles), 59–71
  challenge for future, 59–71
  comparison of Prius and, 63
  comparison of Prius and FCHV, 63
  forecast of fuel cell market introduction, 71
  hybrid system characteristics, 64
  hydrogen production cost, 70
  hydrogen storage technology, 67
  issues for market introduction, 65
  JHFC (Japan Hydrogen and Fuel Cell Demonstration Project), 70
  low temperature performance, 66
  overcoming barriers to commercialization, 67–71
  schematic of hydrogen infrastructure pathways, 68
  Toyota FCHV fuel cell vehicle, 61–62
  Toyota FCVs in Japan and United States, 62
  Toyota fuel cell technology, 61
  Toyota's pathways for eco car, 60
  well-to-wheel CO_2 emissions, 69
  well-to-wheel efficiency, 65
FCV (fuel cell vehicle) markets, future for, 33–58
  buying FCVs, 49–54
  cautionary , but motivational , tale, 55
  energy infrastructure, 39–41
  evoking the future, 48–49
  future for hydrogen and FCVs, 56
  history and future of mobility, 36–37
  information infrastructure, 42–46
  integrating infrastructures, 46–48
  mobility infrastructure, 37–39
  policy goals and social marketing, 54–55
FCVs (fuel cell vehicles), 2, 33, 135, 192
  challenges for, 208–209
  and fuel infrastructure performance and design, 52
  future for hydrogen and, 56
  offer unique bundle of attributes, 52
FCVs (fuel cell vehicles), buying, 49–54
  competing automotive alternatives, 51–52
  driving range and nontravel use of energy, 52–54
  FCV and fuel infrastructure performance and design, 52
  FCVs offer unique bundle of attributes, 52
  new product, 50–51
FCVs in Japan and United States, Toyota, 62
FFVs (flexible-fuel vehicles), 171, 193
Forecast of fuel cell market introduction, 71
FP (Framework Program), 14, 156
Free range transportation, 28–29
Fuel
  chains, 159
  hydrogen as viable alternative, 142–143
  quality, 174
  from renewable resources, 129–131
Fuel availability, effective costs of retail, 196
Fuel cell focus, premature hydrogen, 213–226
  evaluating strategic choices, 217–223
  policy paradigm shift, 215–216
  rationales for change, 216–217
Fuel cell market introduction, forecast of, 71
Fuel cell technology
  industry support for, 143–144
  Toyota, 61
Fuel cell vehicles
  challenges for, 208–209
  hydrogen powered, 188–189
  Toyota FCHV, 61–62
Fuel infrastructure performance and design, FCV, 52
Fuel markets, vehicle and, 177
Fuel vehicles, alcohol, 171
Fuels
  economies of electrolytic hydrogen as motor, 125
  key policy initiatives and alternative, 167–168

Fuels *(Continued)*
low density, 174
new, 199–200
social attributes of new alternative, 176
Fuels, alternative, 165–179
alternative fuels experience is valuable, 165–166
effort of United States, 168–172
hydrogen transition and alternative fuel programs, 166–167
key policy initiatives and alternative fuels, 167–168
lessons learned, 172–178
Fuels and vehicles, transition to new, 191–212
Fuels experience, 15 years of alternative, 165–179
Future
back from, 21–32
hydrogen as viable alternative fuel for, 142–143
representing, 36–37
Future for FCV (fuel cell vehicle) markets, 33–58
buying FCVs, 49–54
cautionary , but motivational , tale, 55
energy infrastructure, 39–41
evoking the future, 48–49
future for hydrogen and FCVs, 56
history and future of mobility, 36–37
information infrastructure, 42–46
integrating infrastructure, 46–48
mobility infrastructure, 37–39
policy goals and social marketing, 54–55
Future for hydrogen and FCVs, 56
Future for hydrogen supply, 88–91
Future, hydrogen in distant, 28–32
continental energy distribution, 29–30
free range transportation, 28–29
transformer technologies, 30–32
Future hydrogen supply options, 74–76
Future hydrogen supply, questions for, 79–84
hypotheses, 82–84
minimizing environmental externalities, 79
observations, 79–82
Future of mobility, history and, 36–37

G

Gasoline, 39–40
Gasoline gallon equivalent (GGE), 193, 203
Generator, electricity, 130
Geographic hydrogen supply issues, 86–87
GGE (gasoline gallon equivalent), 193, 203
GHG emissions, well-to-wheel, 81
GHG (greenhouse gas), 55, 219
Global hydrogen production in 1999, 80
Global warming, 219–221
Goals
percent of sales of AFVs and EPACT, 169
policy, 54–55
Government policies, 13–15
Greenhouse gas (GHG), 55, 219
Groups, special interest, 174

H

HC (hydrocarbon), 218
Health, public, 217–219
HEVs (hybrid electric vehicles), 10, 51, 152, 192, 233
High Level Group on Hydrogen and Fuel Cell Technologies (HLG), 157
HLG (High Level Group on Hydrogen and Fuel Cell Technologies), 157
Hybrid electric vehicles (HEVs), 10, 233
Hybrid system characteristics, 64
Hybrid vehicles, analysis and experience of, 191–212
Hydrocarbon (HC), 218
Hydrogen
as battery alternative, 127–129
better options and transition to, 153
case for, 236
clean, 93–104
coal-derived, 95–97
and electricity, 23–24
end game, 8–9
energy arbitrage from electrolytic, 132
entrepreneurship, 11–13
powered fuel cell vehicles, 188–189
production, 93–95
production cost, 70
rushing headlong toward, 148–150
solar thermal, 130
synergies of, 87–88

Hydrogen *(Continued)*
technology change and transition to, 206–208
technology validation cycle for, 142
as viable alternative fuel for future, 142–143
Hydrogen activities, U.S., 155–164
comparing both approaches, 158–162
European perspective, 155–164
history of EU commitment, 156–158
history of U.S. hydrogen commitment, 155–156
Hydrogen age, strategies for, 21–32
architecture of energy system, 25–27
hydrogen in distant future, 28–32
reasons for using hydrogen, 22–23
Hydrogen and FCVs, future for, 56
Hydrogen as motor fuel, economies of electrolytic, 125
Hydrogen commitment, history of U.S., 155–156
Hydrogen demand for vehicles, scenarios for, 83
Hydrogen economy, international partnership for, 144–145
Hydrogen electrolysis, efficiencies of, 124
Hydrogen from electrolysis, 121–133
electrolysis as energy arbitrage, 131–133
fuel from renewable resources, 129–131
hydrogen as battery alternative, 127–129
PEM electrolysis, 123–126
power quality, 127–129
Proton Energy Systems, 121–123
today's industrial hydrogen products, 126–127
Hydrogen fuel cell focus, premature, 213–226
Hydrogen hope or hype, 235–239
case for hydrogen, 236
cost and availability challenges, 236–238
precarious situation, 238–239
Hydrogen in distant future, 28–32
continental energy distribution, 29–30
free range transportation, 28–29
transformer technologies, 30–32
Hydrogen infrastructure pathways, schematic of, 68
Hydrogen initiative, President's U.S., 135–146
DOE's R&D strategy, 139–142
hydrogen as viable alternative fuel for future, 142–143
industry support for fuel cell technology, 143–144
industry support for hydrogen technology, 143–144
International Partnership for the Hydrogen Economy (IPHE), 144–145
timing of hydrogen transition, 136–139
Hydrogen production in 1999, global, 80
Hydrogen products, today's industrial, 126–127
Hydrogen, reasons for using, 22–23
hydrogen and electricity, 23–24
Hydrogen, sources of, 73–92
challenges facing long term supply options, 76–78
designing hydrogen supply system, 78
existing energy system, 87–88
future for hydrogen supply, 88–91
future hydrogen supply options, 74–76
geographic and regional hydrogen supply issues, 86–87
low cost hydrogen transition, 84–86
matching supply and demand, 84–86
questions for future hydrogen supply, 79–84
synergies of hydrogen, 87–88
Hydrogen storage technology, 67
Hydrogen supply, future for, 88–91
Hydrogen supply issues, geographic and regional, 86–87
Hydrogen supply options, future, 74–76
Hydrogen supply, questions for future, 79–84
hypotheses, 82–84
minimizing environmental externalities, 79
observations, 79–82
Hydrogen supply, system considerations and, 73–92
challenges facing long term supply options, 76–78
designing hydrogen supply system, 78
existing energy system, 87–88
future for hydrogen supply, 88–91
future hydrogen supply options, 74–76

Hydrogen supply *(Continued)*
geographic and regional hydrogen supply issues, 86–87
low cost hydrogen transition, 84–86
matching supply and demand, 84–86
questions for future hydrogen supply, 79–84
synergies of hydrogen, 87–88
Hydrogen supply system, designing, 78
Hydrogen technology, industry support for, 143–144
Hydrogen today, alternatives to, 17–18
Hydrogen transition, 147–154
and alternative fuel programs, 166–167
better options and transition to hydrogen, 153
California perspective, 147–154
getting to and through, 151–152
getting to and through hydrogen transition, 151–152
low cost, 84–86
rushing headlong toward hydrogen, 148–150
strategies for, 9–11
timing of, 136–139
in United States, 137
what was learned about past programs, 152–153
Hydrogen transition, entrepreneurship in, 105–119
business model, 115–117
current policy focus, 105–106
enabling technologies, 114
entrepreneurs as agents of change, 112–113
how problem, 106–107
implications for energy policies, 117–118
innovation in assembled products, 110–111
innovation in nonassembled products, 111–112
innovation perspective, 108–110
service concept for technology, 107–108
starting in niche markets, 113–114
transportation, 105–106
Hydrogen transportation, challenge of, 1–4

I

ICE (internal combustion engines), 3, 48, 184, 233
ICEVs (ICE-powered vehicles), 50
IGCC (integrated gasification combined-cycle), 97
ILEVs (inherently low emission vehicles), 183
Imports, growth in U.S. net oil, 5
Industrial hydrogen products, today's, 126–127
Industries, coordination between auto and oil, 176
Industry support
for fuel cell technology, 143–144
for hydrogen technology, 143–144
Information infrastructure, 42–46
Internet, 43–46
telephony, 42–43
wireless networking, 46
Infrastructure and market development, building, 175
Infrastructure, energy, 39–41
electricity, 40–41
gasoline, 39–40
Infrastructure, information, 42–46
Internet, 43–46
telephony, 42–43
wireless networking, 46
Infrastructure, mobility, 37–39
roads, 37–38
vehicles, 38–39
Infrastructure pathways, schematic of hydrogen, 68
Infrastructures, integrating, 46–48
automobility and communications, 46–47
mobile electricity, 47–48
Inherently low emission vehicles (ILEVs), 183
Initiatives, key policy, 167–168
Innovation
in assembled products, 110–111
in nonassembled products, 111–112
perspective, 108–110
Integrated gasification combined-cycle (IGCC), 97
Interest groups, special, 174
Internal combustion engines (ICE), 3, 48, 184, 233
International Partnership for the Hydrogen Economy (IPHE), 144–145, 156

Internet, 43–46
Introduction
forecast of fuel cell market, 71
issues for market, 65
Investments, private sector, 175
Investors, attracting, 177
IP (Internet protocol), 44
IPHE (International Partnership for the Hydrogen Economy), 144–145, 156
ITS-Davis (Institute of Transportation Studies of U. C.-Davis), 7

J

Japan Hydrogen and Fuel Cell Demonstration Project (JHFC), 70
Japan, Toyota FCVs in, 62
JHFC (Japan Hydrogen and Fuel Cell Demonstration Project), 69–70

K

Key policy initiatives and alternative fuels, 167–168
KW (kilowatt), 31

L

Land, global mass of vertebrate life on, 230
LBD (learning-by-doing), 198–199
Learning, 202
Learning-by-doing (LBD), 198–199
LEVs (low emission vehicles), 183
LHV (lower heating value), 98
Life, global mass of vertebrate, 230
Light duty diesel vehicles, 186–188
Light duty market, 177
Limited diversity, effective cost of, 197–198
Liquefied natural gas (LNG), 95, 166
Liquefied petroleum gas (LPG), 166
Lithium batteries, 231–233
LNG (liquefied natural gas), 95, 166
Long-run barrier, 204–206
Long term supply options, challenges facing, 76–78
Low cost hydrogen transition, 84–86
Low density fuels, 174
Low emission vehicles (LEVs), 183
Low temperature performance, 66
Lower heating value (LHV), 98
LPG (liquefied petroleum gas), 166

M

M85, 183–184
Manufacturers' costs per model, vehicle, 197
Market development, building infrastructure and, 175
Market factors, 223
Market introduction
forecast of fuel cell, 71
issues for, 65
Market segment, small, 202–204
Market values, transport vs. stationary, 129
Marketing, social, 54–55
Markets
AFV penetration in, 178
consumer, 174
light duty, 177
niche, 173
starting in niche, 113–114
vehicle and fuel, 177
Markets, future for FCV (fuel cell vehicle), 33–58
buying FCVs, 49–54
cautionary , but motivational , tale, 55
energy infrastructure, 39–41
evoking the future, 48–49
future for hydrogen and FCVs, 56
history and future of mobility, 36–37
information infrastructure, 42–46
integrating infrastructures, 46–48
mobility infrastructure, 37–39
policy goals and social marketing, 54–55
Megawatt (MW), 97
Methanol, 170–171
METI (Ministry of Economy , Trade and Industry), 69
Mobile electricity, 47–48
Mobility; *See also* Automobility
Mobility, history and future of, 36–37
Mobility infrastructure, 37–39
roads, 37–38
vehicles, 38–39
Model
business, 115–117
vehicle manufacturers' costs per, 197
Model diversity, accounting for vehicle, 197–198
Motor fuels, economies of electrolytic hydrogen as, 125

MTBE (methyl tertiary butyl ether), 170
MW (megawatt), 97

N

National Energy Policy Development Group (NEPDG), 155
National energy policy, holistic approach to, 181–182
National Energy Policy Plan (NEPP), 155
National Energy Technology Laboratory (NETL), 101
Natural gas, steam reforming of, 93–95
Neighborhood electric vehicles (NEVs), 186
NEPDG (National Energy Policy Development Group), 155
NEPP (National Energy Policy Plan), 155
NETL (National Energy Technology Laboratory), 101
Networking, wireless, 46
NEVs (neighborhood electric vehicles), 186
New alternative fuels, social attributes of, 176
New fuels and vehicles, transition to, 191, 199
Niche markets, 173
  starting in, 113–114
Nickelmetal hydride (NiMH), 227
NiMH (nickelmetal hydride), 227
1990s , lessons of, 15–17
Nonassembled products, innovation in, 111–112
NO_X (nitrogen oxide), 218

O

Oil imports, growth in U.S. net, 5
Oil industries, coordination between auto and, 176
Oil prices, importance of, 204–206
Oil production and domestic use, U.S., 6
OPEC (Organization of Petroleum Exporting Countries), 6

P

Partnership, international, 144–145
Parts per million by volume (ppmv), 27
Past programs, what was learned about, 152–153
Pathways
  electrolysis technology, 122
  schematic of hydrogen infrastructure, 68
PCs (personal computers), 110
PEM electrolysis, 123–126
PEM (proton exchange membrane), 121
Performance and design, FCV and fuel infrastructure, 52
Performance, low temperature, 66
Personal computers (PCs), 110
Perspective, innovation, 108–110
Petroleum demand, reducing, 192–193
Petroleum, steam reforming of, 93–95
Policies
  government, 13–15
  holistic approach to national energy, 181–182
  implications for energy, 117–118
Policy goals and social marketing, 54–55
Policy initiatives, key, 167–168
Policy paradigm shift, 215–216
Power quality, 127–129
Power, sustainable, 132
PPIC (Public Policy Institute of California), 148
Ppmv (parts per million by volume), 27
Premature hydrogen fuel cell focus, 213–226
President's U.S. hydrogen initiative, 135–146
  DOE's R&D strategy, 139–142
  hydrogen as viable alternative fuel for future, 142–143
  industry support for fuel cell technology, 143–144
  industry support for hydrogen technology, 143–144
  International Partnership for the Hydrogen Economy (IPHE), 144–145
  timing of hydrogen transition, 136–139
Prices, importance of oil, 204–206
Prius and FCHV, comparison of, 63
Private sector investment, 175
Problem, chicken or egg, 213–226
Production and domestic use, U.S. oil, 6
Production cost, hydrogen, 70
Production, hydrogen, 93–95

Products
  innovation in assembled, 110–111
  innovation in nonassembled, 111–112
  today's industrial hydrogen, 126–127
Programs
  hydrogen transition and alternative fuel, 166–167
  what was learned about past, 152–153
Proton Energy Systems, 121–123
Proton exchange membrane (PEM), 121
Public health, 217–219
Public Policy Institute of California (PPIC), 148
PUCs (public utility commissions), 175

Q

Quality
  fuel, 174
  power, 127–129

R

R&D (research and development), 4, 136
R&D strategy, DOE's, 139–142
RD&D (research, development, and demonstration), 192
Reduction, technological cost, 198–199
Regional hydrogen supply issues, 86–87
Renewable resources, fuel from, 129–131
Renewable sources, sustainable power from, 132
Resources, fuel from renewable, 129–131
Retail fuel availability, effective costs of, 196
Roads, 37–38

S

Scale, 202
  economies of, 197
SCF (standard cubic feet), 127
Security, energy, 4–7, 221–222
Sequestration
  carbon taxes and CO_2, 102–103
  CO_2, 101–102
  CO_2 capture and, 93–104
Service concept for technology, 107–108
Small market segment and buying AFVs, 202–204
SMR (steam methane reforming), 31
Social attributes of new alternative fuels, 176
Social marketing, policy goals and, 54–55
Solar energy supply, 229
Solar thermal hydrogen and electricity generator, 130
Sources, sustainable power from renewable, 132
Special interest groups, 174
Standard cubic feet (SCF), 127
Stationary market values, transport vs., 129
Steam methane reforming (SMR), 31
Steam reforming of natural gas and petroleum, 93–95
Storage technology, hydrogen, 67
Strategic choices, evaluating, 217–223
  energy security, 221–222
  global warming, 219–221
  market factors, 223
  public health, 217–219
Strategies
  DOE's R&D, 139–142
  for hydrogen age, 21–32
SULEVs (super ultra low emission vehicles), 183
Supplies
  future for hydrogen, 88–91
  questions for future hydrogen, 79–84
Supply and demand, matching, 84–86
Supply issues, geographic and regional, 86–87
Supply options
  challenges facing long term, 76–78
  future hydrogen, 74–76
Supply, system considerations and hydrogen, 73–92
Supply system, designing hydrogen, 78
Sustainable power from renewable sources, 132
System characteristics, hybrid, 64
Systems
  architecture of energy, 25–27
  considerations and hydrogen supply, 73–92
  designing hydrogen supply, 78
  existing energy, 87–88

T

TAFV (transitional alternative fuel and vehicle model), 194
Tales, cautionary , but motivational, 55
Taxes, carbon, 102–103
Technological cost reduction, 198–199
Technologies
  changes and transitions to hydrogen, 206–208
  conventional, 206
  enabling, 114
  hydrogen storage, 67
  industry support for fuel cell, 143–144
  industry support for hydrogen, 143–144
  service concepts for, 107–108
  Toyota fuel cell, 61
  transformer, 30–32
  transition, 177
  validation cycles for hydrogen, 142
Technology pathways, electrolysis, 122
Telephony, 42–43
Temperature, low, 66
Thermal hydrogen, solar, 130
Toyota FCHV fuel cell vehicle, 61–62
Toyota FCVs in Japan and United States, 62
Toyota fuel cell technology, 61
Toyota's pathways for eco car, 60
Transformer technologies, 30–32
Transition
  conventional technology and forestalling, 206
  entrepreneurship in hydrogen, 105–119
  getting to and through hydrogen, 151–152
  hydrogen, 166–167
  to hydrogen, 206–208
  low cost hydrogen, 84–86
  to new fuels and vehicles, 191, 199
  strategies for hydrogen, 9–11
  technologies, 177
  timing of hydrogen, 136–139
Transitional alternative fuel and vehicle model (TAFV), 194
Transitional analysis, 193–195
Transitions matter a lot, 200–202
Transport vs. stationary market values, 129
Transportation
  challenge of hydrogen, 1–4
  CO_2, 99–101
  free range, 28–29
  reducing petroleum demand in, 192–193

U

U.S. hydrogen activities, 155–164
  comparing both approaches, 158–162
  European perspective, 155–164
  history of EU commitment, 156–158
  history of U.S. hydrogen commitment, 155–156
U.S. hydrogen commitment, history of, 155–156
U.S. hydrogen initiative, President's, 135–146
U.S. net oil imports, growth in, 5
U.S. oil production and domestic use, 6
ULEVs (ultra low emission vehicles), 183
United States, Toyota FCVs in, 62

V

Validation cycle, technology, 142
Vehicle and fuel markets, 177
Vehicle emissions, lower, 170
Vehicle manufacturers' costs per model, 197
Vehicle model diversity, accounting for, 197–198
Vehicles, 38–39
  alcohol fuel, 171
  analysis and experience of alternative, 191–212
  analysis and experience of hybrid, 191–212
  challenges for fuel cell, 208–209
  hydrogen powered fuel cell, 188–189
  light duty diesel, 186–188
  scenarios for hydrogen demand for, 83
  Toyota FCHV fuel cell, 61–62
  transition to new fuels and, 191, 199
Vertebrate life, global mass of, 230
Viable alternative fuel, hydrogen as, 142–143

W

Warming, global, 219–221
Well-to-wheels
  air pollutant emissions, 81
  efficiency, 65
  GHG emissions, 81
Wh/kg (Watt hours per kilogram), 227
Wireless networking, 46

Z

ZEVs (zero emission vehicles), 13, 170, 185, 206, 217
ZLEVs (zero level emissions vehicles), 219